Networks

F. R. Connor

Ph.D., M.Sc., B.Sc.(Eng.) Hons, A.C.G.I.,
C.Eng., M.I.E.E., M.I.E.R.E., M.Inst.P.

Edward Arnold

First published 1972 by Edward Arnold (Publishers) Ltd., 25 Hill Street,
London W1X 8LL

Reprinted 1975

ISBN 0 7131 3258 2

Photoset by The Universities Press, Belfast
and printed in Great Britain by The Pitman Press, Bath.

Preface

This is an introductory book on the important topic of Networks. Such networks are used extensively in the fields of Electronics and Tele-communications, and the book endeavours to present the basic ideas in a concise and coherent manner. Moreover, to assist in the assimilation of these basic ideas, many worked examples from past examination papers have been provided to illustrate clearly, the application of the fundamental theory.

The first part of the book is devoted to an analysis of the various types of one-port and two-port networks, emphasis being placed on their particular characteristics. Subsequent chapters then deal specifically with the important class of networks known as filters and consideration is given to the practical problem of their design. Finally, the book ends with a chapter on the more modern approach of synthesis, which is applied to the field of filter design, thereby presenting a unified and broad outline of this extremely useful topic.

The book will be found useful by students preparing for London University examinations, degrees of the Council of National Academic Awards, examinations of the Council of Engineering Institutions and for other qualifications such as Higher National Certificates, Higher National Diploma and certain examinations of the City & Guilds of London Institute. It will also be useful to practising engineers in industry who require a ready source of basic knowledge to help them in their applied work.

Acknowledgements

The author wishes to thank the Senate of the University of London and the Council of the Institution of Electrical Engineers, for permission to include questions from their past examination papers. The solutions provided are his own and he accepts full responsibility for them.

Finally, the author would like to thank the publishers for various useful suggestions and will be grateful to his readers, for drawing his attention to any errors which may have occurred.

F.R.C.

Books in this series by the same author:

1. Signals

In preparation:

3. Wave transmission
4. Antennas
5. Modulation
6. Noise

Contents

Symbols used in the book

ω angular frequency
ω_r or ω_0 frequency at resonance
Y admittance $\quad \left. \begin{matrix} Z_0 \\ Z_{0T} \\ Z_{0\pi} \end{matrix} \right\}$ Z_0 characteristic impedance
Z impedance Z_{0T} characteristic impedance
X reactance $Z_{0\pi}$ of T- or π-section
f_c cut-off frequency
f_∞ frequency of infinite attenuation
g conductance parameter
y admittance parameter
Q_0 Q-factor
α attenuation coefficient
β phase change coefficient
γ propagation coefficient

Abbreviations used in the book

I.E.E.   The Institution of Electrical Engineers
L.U.   University of London, BSc(Eng)
BSc(Eng) Tels.   examination in Telecommunications, Part 3

1
Introduction

Networks are used extensively in telecommunication circuits and form the basic structure of the transmission line or composite filter.

At present, the most widely used network is the *passive* network which has no internal voltages. A passive network is therefore an arrangement of components such as resistors, inductors or capacitors in the form of a physical structure; various types of structure are possible, each having its own particular properties and applications.

However, more recently, *active* networks that contain devices such as valves or transistors, are being used more often, as they possess some useful advantages over the usual passive networks.

1.1 Passive networks

The two most basic structures are the one-port network and the two-port network shown in Fig. 1. A one-port network has a pair of terminals only, while a two-port network has two input terminals and two output terminals.

It is common practice to regard the network as a 'black-box' to which the appropriate terminals are attached and an analysis is made on this basic 'black-box' configuration.

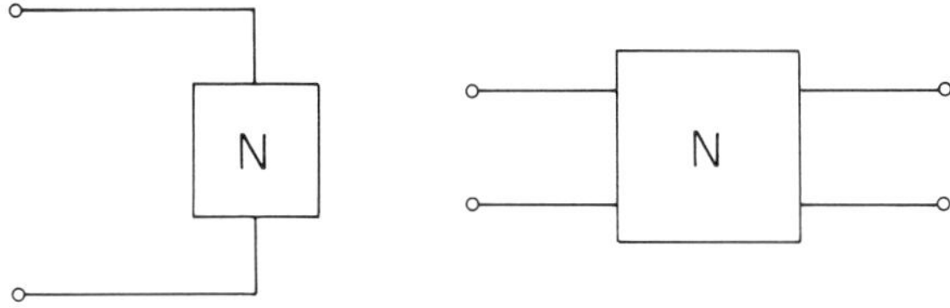

Fig. 1 One-port network (*left*) and two-port network

1.2 Examples of networks

Typical networks used in communication systems are shown in block schematic in Fig. 2.

2 Networks

Networks are generally classified as *symmetrical* or *asymmetrical* and are either of the *balanced* or *unbalanced* forms. Symmetrical networks are unchanged when input and output ports are interchanged whereas asymmetrical networks are changed. Networks are 'balanced' when their components are correctly placed with respect to an 'earth plane' so that stray capacitances balance out and produce less interference or 'crosstalk' between circuits.

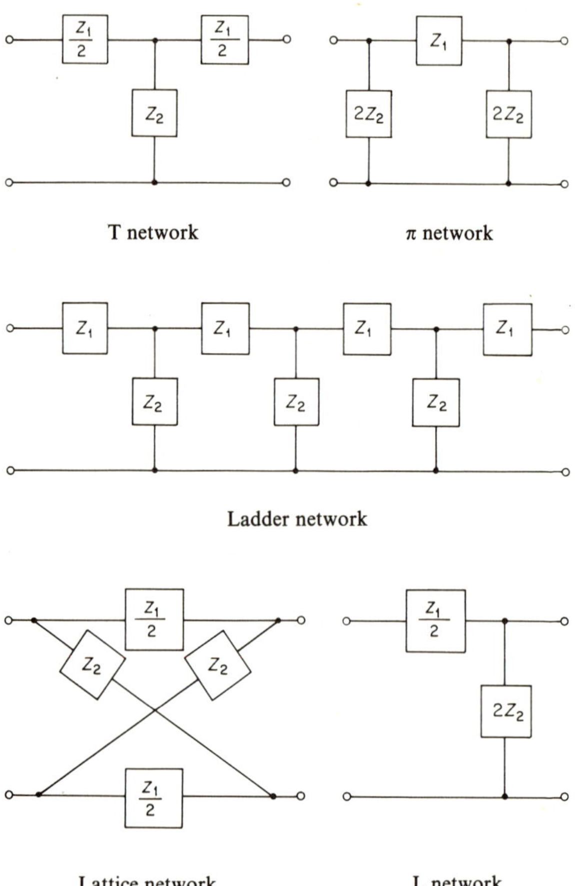

Fig. 2

1.3 Properties of networks

The transmission properties of networks are best described in terms of the attenuation and phase-change coefficients of a network, both of which are functions of frequency.

In addition, it is also necessary to make a more detailed study of the impedances of a network, which have a profound effect on its transmission properties, since generally the telecommunication network consists of inductors and capacitors whose impedances vary with frequency.

A network is therefore characterised by its *iterative* and *image* impedances, which are generally different. The former are important for 'matching' one network to another, while the latter determine the maximum transfer of power through the network.

1.4 Analysis of networks

Assuming that the network contains only linear passive components, the analysis can be done by considering the voltages and currents of the network or its impedances. A set of equations can be established between these quantities and it leads to the basic $A\ B\ C\ D$ parameters* which are well known in power and transistor circuits.

In telecommunication circuits, however, one is more concerned with the transmission properties of the network as a function of frequency. It is usual, therefore, to treat the network in such a way as to bring out these properties in the analysis.

In the case of lossless one-port networks, this will be done by considering the *poles* and *zeros* of the network in the light of Foster's reactance theorem. For two-port networks, usual algebraic methods will be employed, though advanced techniques also employ the concept of poles and zeros, which leads to *modern network theory*; this will be applied to filters in Chapter 6.

1.5 Active networks

As mentioned earlier, these networks are becoming more popular in certain applications, especially at the lower frequencies. One advantage of them is that heavy and costly inductors can be eliminated and replaced by *RC* networks, through the use of operational amplifiers, gyrators or

---

* See Appendix C.

negative impedance convertors. Moreover, by using active devices such as transistors, insertion loss may be compensated for through the network, in addition to obtaining the desired response characteristic.

A study of such networks is a subject in its own right and will be dealt with briefly in Chapter 6.

2
One-port networks

The two most important examples are the series tuned circuit and the parallel tuned circuit.

2.1 Series tuned circuit

Let Z be the input impedance of a series circuit consisting of R, L and C, connected to a generator of variable frequency, as shown in Fig. 3.

We have

$$Z = R + j\left(\omega L - \frac{1}{\omega C}\right) = \frac{\omega CR + j(\omega^2 LC - 1)}{\omega C} \tag{1}$$

At resonance, $\omega = \omega_r$ and so

$$\omega_r L = \frac{1}{\omega_r C}$$

or

$$\omega_r = \frac{1}{\sqrt{LC}} \tag{2}$$

and

$$f_r = \frac{1}{2\pi\sqrt{LC}}$$

From equations (1) and (2) we obtain when $\omega = \omega_r$

$$Z = R, \text{ which is a pure resistance at resonance.}$$

In general, equation (1) can be written as

$$Z = \frac{\omega CR + jLC[\omega^2 - (1/LC)]}{\omega C} = R + j\frac{L}{\omega}(\omega^2 - \omega_r^2) \text{ since } \omega_r^2 = \frac{1}{LC}$$

For a lossless network, $R = 0$, giving

$$Z = j\frac{L}{\omega}(\omega^2 - \omega_r^2) \tag{3}$$

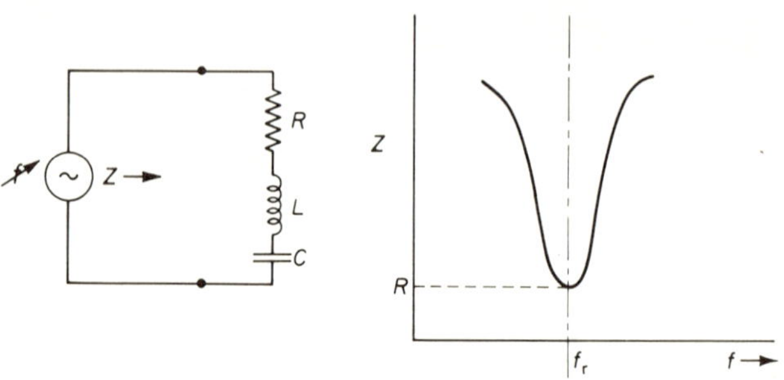

Fig. 3

2.2   Parallel tuned circuit

Let Z be the input impedance of a parallel circuit consisting of L, R and C, where R is in series with the coil and is shown in Fig. 4.

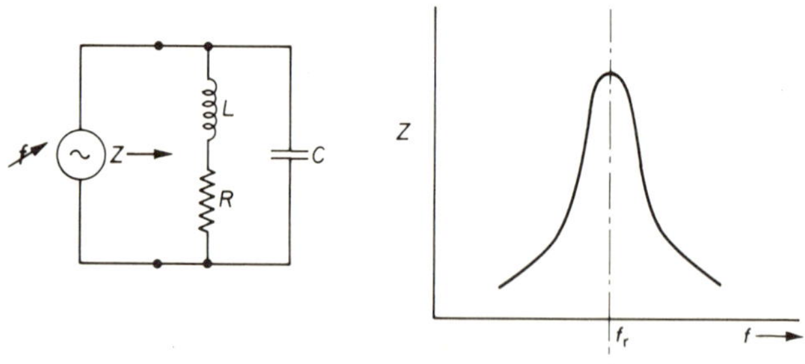

Fig. 4

We have

$$Z = \frac{(R + j\omega L)[-j(1/\omega C)]}{R + j[\omega L - (1/\omega C)]} = \frac{R + j\omega L}{j\omega CR + (1 - \omega^2 LC)}$$

Usually $\omega L \gg R$ if the Q of the coil circuit is large where

$$Q = \frac{\omega L}{R}$$

Hence
$$Z = \frac{j\omega L}{j\omega CR + (1 - \omega^2 LC)} \tag{4}$$

At resonance $\omega = \omega_r$ and

$$\omega_r L = \frac{1}{\omega_r C}$$

or

$$\omega_r = \frac{1}{\sqrt{LC}}$$

and

$$f_r = \frac{1}{2\pi\sqrt{LC}}$$

Hence

$$Z = \frac{j\omega_r L}{j\omega_r CR} = \frac{L}{CR} = Z_d$$

which is called the *dynamic impedance* of the circuit and is a pure resistance.

In general, equation (4) can be written as

$$Z = \frac{j\omega L}{j\omega CR + LC\left(\dfrac{1}{LC} - \omega^2\right)} = \frac{j\omega L}{j\omega CR + LC(\omega_r^2 - \omega^2)}$$

For a lossless network, $R = 0$ giving

$$Z = \frac{j\omega L}{LC(\omega_r^2 - \omega^2)} = \frac{j\omega}{C}\left[\frac{1}{\omega_r^2 - \omega^2}\right]$$

or

$$Z = \frac{\omega}{jC}\left[\frac{1}{\omega^2 - \omega_r^2}\right] \tag{6}$$

Again, from equation (4) we have

$$Z = \frac{j\omega L}{j\omega CR + (1 - \omega^2 LC)}$$

$$= \frac{j\omega L}{j\omega CR\left[1 + \dfrac{LC}{j\omega CR}\left(\dfrac{1}{LC} - \omega^2\right)\right]} = \frac{\dfrac{L}{CR}}{\left[1 + \dfrac{LC}{j\omega CR}\left(\dfrac{1}{LC} - \omega^2\right)\right]}$$

Since

$$Q \simeq \frac{\omega L}{R} \simeq \frac{1}{\omega CR}$$

when ω is *close* to ω_r and

$$LC \simeq \frac{1}{\omega_r^2}$$

Hence

$$Z = \frac{L/CR}{1 + (Q/j\omega_r^2)(\omega_r^2 - \omega^2)} = \frac{Z_d}{1 + j(Q/\omega_r^2)(\omega^2 - \omega_r^2)}$$

Now $\omega^2 = (\omega_r \pm \delta\omega)^2 \simeq \omega_r^2 \pm 2\omega_r . \delta\omega$ since $(\delta\omega)^2$ is very small for values of ω *close* to resonance.

Hence $(\omega^2 - \omega_r^2) = 2\omega_r \, \delta\omega$ (choosing the positive value)

and

$$Z = \frac{Z_d}{1 + 2j\dfrac{Q\delta\omega}{\omega_r}} = \frac{Z_d}{1 + 2jQ\delta}$$

where

$$\delta = \frac{\delta\omega}{\omega_r} = \frac{\delta f}{f_r}$$

or

$$|Z| = \frac{Z_d}{\sqrt{1 + 4Q^2 . \delta^2}} = \frac{L/CR}{\sqrt{1 + 4Q^2 . \delta^2}}$$

which gives the general impedance Z *close* to resonance, in terms of Z_d, Q and δ.

EXAMPLE 1

A constant-voltage generator is connected to a parallel-tuned circuit consisting of a coil of inductance 250 μH and resistance 12·5 Ω in parallel with a variable capacitor. If the circuit is resonant at a frequency of 1 MHz, calculate its impedance at 4 kHz off resonance.

Solution
We have

$$f_r = \frac{1}{2\pi\sqrt{LC}}$$

or

$$C = \frac{1}{4\pi^2 L f_r^2} = \frac{1}{4\pi^2 . 250 . 10^{-6} . 10^{12}} \simeq 100 \text{ pf}$$

Now

$$Q = \omega_r \frac{L}{R} = \frac{2\pi . 10^6 . 250 . 10^{-6}}{12 \cdot 5} = 40\pi \simeq 125$$

Also
$$Z_\mathrm{d} = \frac{L}{CR} = \frac{250 \cdot 10^{-6}}{100 \cdot 10^{-12} \cdot 12 \cdot 5} = 200 \text{ k}\Omega$$

and
$$Z = \frac{Z_\mathrm{d}}{\sqrt{1 + 4Q^2\delta^2}}$$

where
$$\delta = \frac{\delta\omega_\mathrm{r}}{\omega_\mathrm{r}} = \frac{\delta f_\mathrm{r}}{f_\mathrm{r}} = \frac{4000}{10^6} = \frac{1}{250}$$

Now
$$\sqrt{1 + 4Q^2\delta^2} = \sqrt{1 + 4 \cdot (125)^2 \cdot \frac{1}{(250)^2}} = \sqrt{2}$$

Hence
$$Z = \frac{200 \text{ k}\Omega}{\sqrt{2}}$$

or
$$Z = 141 \text{ k}\Omega$$

2.3 Lossless networks

The input impedance Z of a one-port lossless network may assume the forms given by equations (3) and (6). These can be combined to give a single general expression for any combination of such networks and the input impedance is of form

$$Z = A\left[\frac{(\omega^2 - \omega_1^2)(\omega^2 - \omega_3^2)\cdots}{(\omega^2 - \omega_2^2)(\omega^2 - \omega_4^2)\cdots}\right] \tag{7}$$

where $A = \mathrm{j}\omega H$ or $H/\mathrm{j}\omega$, H being a scale factor.

An alternative form for Z can be obtained by the substitution $s = \sigma + \mathrm{j}\omega$, with $\sigma = 0$ for a lossless network. Hence we obtain,

$$Z = A\left[\frac{(s^2 + \omega_1^2)(s^2 + \omega_3^2)\cdots}{(s^2 + \omega_2^2)(s^2 + \omega_4^2)\cdots}\right] \tag{8}$$

where $A = sH$ or H/s, H being a scale factor.

2.4 Poles and zeros

The values of ω which make $Z = 0$ in (7) such as ω_1, ω_3 etc. are called the *zeros* of Z and are denoted by a small circle 'O'. Similarly, the values of ω which make $Z = \infty$ such as ω_2, ω_4 etc. are called the *poles* of Z and are denoted by a cross 'x'.

Poles and zeros can be plotted along the X-axis, with reactance $\pm jX$ along the Y-axis in the ω-plane or alternatively, in the complex s-plane, as shown in Fig. 5. The concept of poles and zeros is of great importance in modern network theory and in the field of control engineering.

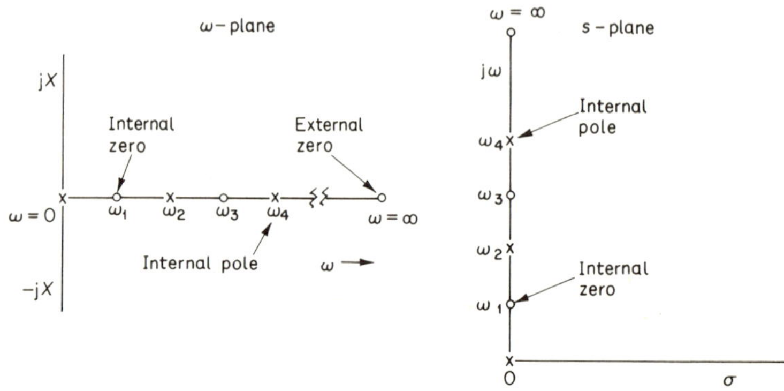

Fig. 5

2.5   Foster's reactance theorem[1-2]

Much information about the behaviour of a lossless network may be obtained from a study of its poles and zeros and this is concisely expressed by Foster's reactance theorem as follows:

The input impedance of any one-port lossless network is completely specified by its internal poles and zeros which occur at real frequencies and by a scale factor H, which is a real constant.

As a result of this theorem, two useful rules may be stated:

Rule 1.   The poles and zeros of a one-port lossless network must alternate along the frequency axis. This is called the separation property.

Rule 2.   Four types of reactance curves may be obtained with a lossless one-port network.

Notes

(a) $\omega = 0$ is called an *external* pole and $\omega = \infty$ is called an *external* zero in Fig. 5.

(b) $\omega_1, \omega_2, \omega_3 \cdots$ etc. are the *internal* poles and zeros and by Rule 1 above, the separation property implies that $\omega_1 < \omega_2 < \omega_3$ etc.

(c) Foster's reactance theorem does not apply to the *external* poles and zeros, but it is useful to consider them first when drawing reactance curves.

(a) Foster networks

The four possible forms depend upon the position of the poles and zeros along the frequency axis (see Figs. 6 and 7).

Form A

$$Z = \mathrm{j}\omega H\left[\frac{(\omega^2 - \omega_2^2)(\omega^2 - \omega_4^2)\cdots(\omega^2 - \omega_{2z}^2)}{(\omega^2 - \omega_1^2)(\omega^2 - \omega_3^2)\cdots(\omega^2 - \omega_{2z+1}^2)}\right]$$

where z is a real number.

Form B

$$Z = \mathrm{j}\omega H\left[\frac{(\omega^2 - \omega_2^2)(\omega^2 - \omega_4^2)\cdots(\omega^2 - \omega_{2z}^2)}{(\omega^2 - \omega_1^2)(\omega^2 - \omega_3^2)\cdots(\omega^2 - \omega_{2z-1}^2)}\right]$$

Form C

$$Z = \frac{H}{\mathrm{j}\omega}\left[\frac{(\omega^2 - \omega_1^2)(\omega^2 - \omega_3^2)\cdots(\omega^2 - \omega_{2p+1}^2)}{(\omega^2 - \omega_2^2)(\omega^2 - \omega_4^2)\cdots(\omega^2 - \omega_{2p}^2)}\right]$$

where p is a real number.

Form D

$$Z = \frac{H}{\mathrm{j}\omega}\left[\frac{(\omega^2 - \omega_1^2)(\omega^2 - \omega_3^2)\cdots(\omega^2 - \omega_{2p-1}^2)}{(\omega^2 - \omega_2^2)(\omega^2 - \omega_4^2)\cdots(\omega^2 - \omega_{2p}^2)}\right]$$

For these expressions, four different series-type networks can be constructed and are called the *canonic* forms or *first* Foster forms. By the use of admittance methods or by the principle of duality, four parallel-type or dual networks can be constructed and are called the *second* Foster forms. The various networks and their associated reactance curves are shown in Figs. 6 and 7. For convenience, assume $z = p = 2$, as this would restrict the number of terms in Z to about two or three.

12   *Networks*

FIRST FOSTER NETWORKS

Type 1

$$Z = j\omega H\left[\frac{(\omega^2 - \omega_2^2)}{(\omega^2 - \omega_1^2)(\omega^2 - \omega_3^2)}\right]$$

Type 2

$$Z = j\omega H\left[\frac{(\omega^2 - \omega_2^2)}{(\omega^2 - \omega_1^2)}\right]$$

Type 3

$$Z = \frac{H}{j\omega}\left[\frac{(\omega^2 - \omega_1^2)(\omega^2 - \omega_3^2)}{(\omega^2 - \omega_2^2)}\right]$$

Type 4

$$Z = \frac{H}{j\omega}\left[\frac{(\omega^2 - \omega_1^2)}{(\omega^2 - \omega_2^2)}\right]$$

SECOND FOSTER NETWORKS

It is more useful to draw reactance curves, though by using admittance methods we could also draw susceptance curves if required.

Type 5

$$Z = \frac{H}{j\omega}\left[\frac{(\omega^2 - \omega_1^2)(\omega^2 - \omega_3^2)}{(\omega^2 - \omega_2^2)}\right]$$

Type 6

$$Z = \frac{H}{j\omega}\left[\frac{(\omega^2 - \omega_1^2)}{(\omega^2 - \omega_2^2)}\right]$$

Type 7

$$Z = j\omega H\left[\frac{(\omega^2 - \omega_2^2)}{(\omega^2 - \omega_1^2)(\omega^2 - \omega_3^2)}\right]$$

Type 8

$$Z = j\omega H\left[\frac{(\omega^2 - \omega_2^2)}{(\omega^2 - \omega_1^2)}\right]$$

jX
-jX
ω₁
ω₂
ω₃
ω
Type 1
L₁
C₁
L₃
C₃
jX
-jX
ω₁
ω₂
ω
Type 2
L₀
L₁
C₁
jX
-jX
ω₁
ω₂
ω₃
ω
Type 3
L₀
C₀
L₂
C₂
jX
-jX
ω₁
ω₂
ω
Type 4
C₀
L₂
C₂

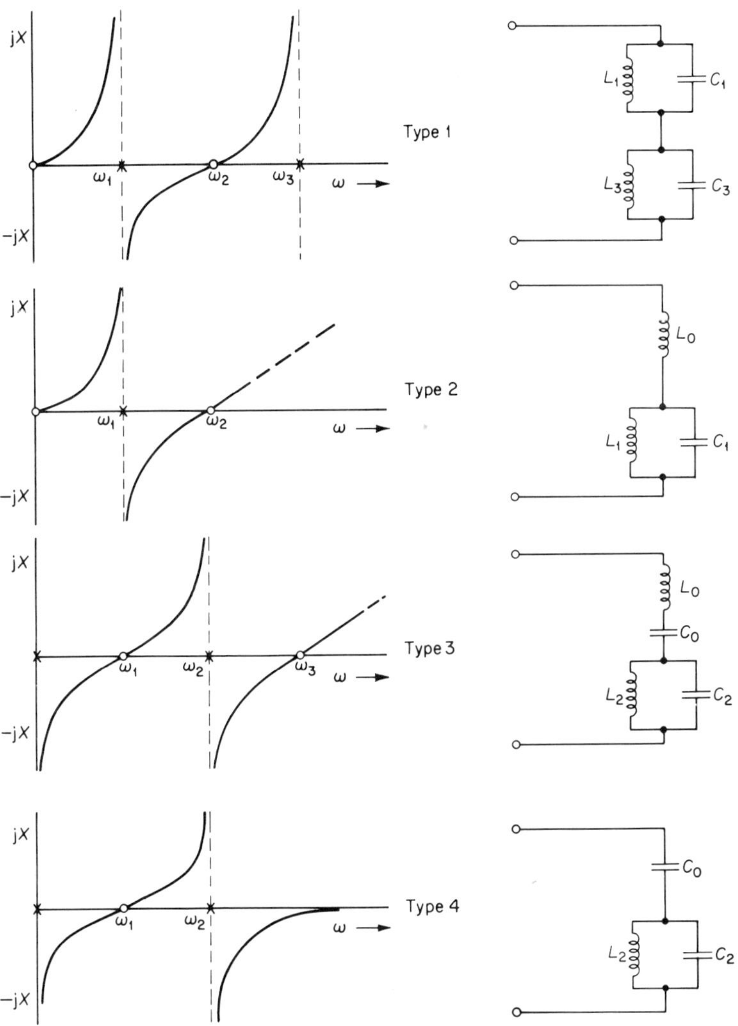

Fig. 6

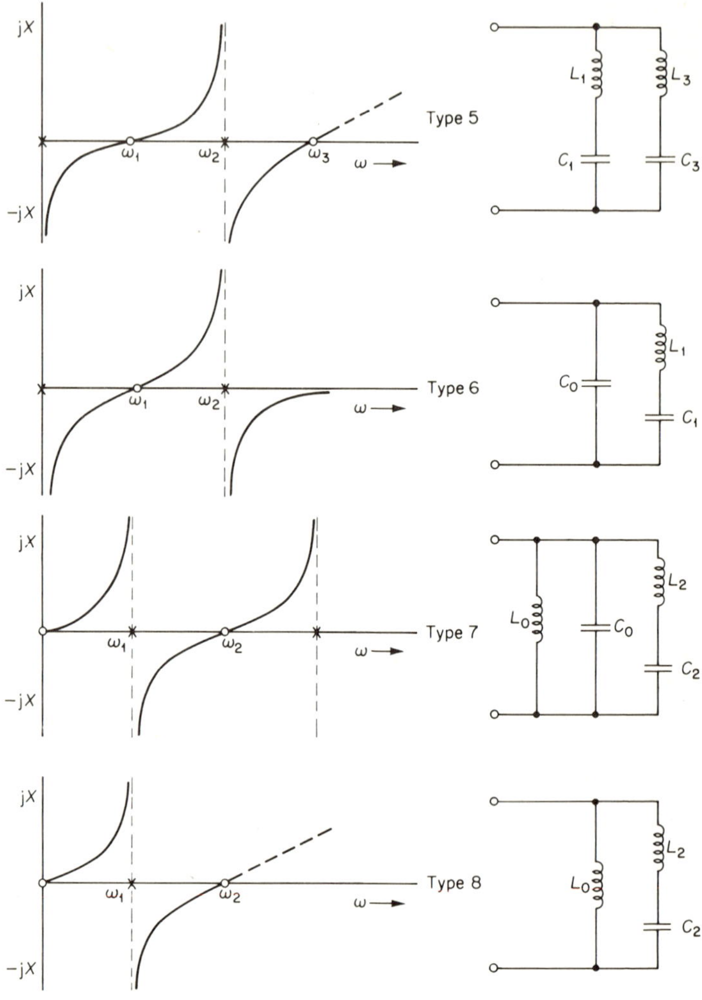

Fig. 7

(b) Synthesis of Foster networks

The synthesis of a network to give a certain reactance curve, can be determined by expanding the equation for Z as partial fractions and identifying each fraction as a physical component. This is illustrated by the following example.

EXAMPLE 2
Synthesize a series Foster network with two internal poles at ω_1 and ω_3 and one internal zero at ω_2 to give the reactance curves of Type 1 in Section 2.5(a).

Solution

Here
$$Z = j\omega H \left[\frac{(\omega^2 - \omega_2^2)}{(\omega^2 - \omega_1^2)(\omega^2 - \omega_3^2)} \right]$$

Let
$$\frac{(\omega^2 - \omega_2^2)}{(\omega^2 - \omega_1^2)(\omega^2 - \omega_3^2)} = \frac{A}{(\omega^2 - \omega_1^2)} + \frac{B}{(\omega^2 - \omega_3^2)}$$

Hence
$$(\omega^2 - \omega_2^2) \equiv A(\omega^2 - \omega_3^2) + B(\omega^2 - \omega_1^2)$$

Putting $\omega = \omega_1$ gives $\omega_1^2 - \omega_2^2 \equiv A(\omega_1^2 - \omega_3^2)$

or
$$A = (\omega_1^2 - \omega_2^2)/(\omega_1^2 - \omega_3^2)$$

Similarly, if $\omega = \omega_3$ gives

$$\omega_3^2 - \omega_2^2 \equiv B(\omega_3^2 - \omega_1^2)$$

or
$$B = \frac{(\omega_3^2 - \omega_2^2)}{(\omega_3^2 - \omega_1^2)}$$

and
$$Z = \frac{j\omega HA}{(\omega^2 - \omega_1^2)} + \frac{j\omega HB}{(\omega^2 - \omega_3^2)}$$

Substituting the values for A and B leads to

$$Z = \frac{j\omega H(\omega_1^2 - \omega_2^2)}{(\omega_1^2 - \omega_3^2)(\omega^2 - \omega_1^2)} + \frac{j\omega H(\omega_3^2 - \omega_2^2)}{(\omega_3^2 - \omega_1^2)(\omega^2 - \omega_3^2)}$$

Comments
 (i) In typical problems usually A and B have numerical values.
(ii) The right-hand side of the equation for Z corresponds to two *series* impedances, each of which can be identified as a *parallel* combination of L and C from Table 1.

   In a typical case the numerical values of ω_1, ω_2 and ω_3 would be given and the scale factor H is evaluated from data specifying the value of Z at some particular ω.

The component values are then obtained with the help of the equations

$$\omega_1^2 = \frac{1}{L_1 C_1} \quad \text{and} \quad \omega_3^2 = \frac{1}{L_3 C_3}$$

The network is shown in Fig. 8.

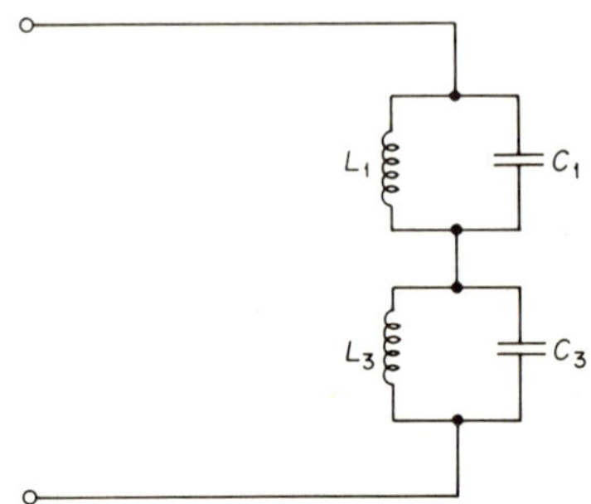

Fig. 8

Table 1

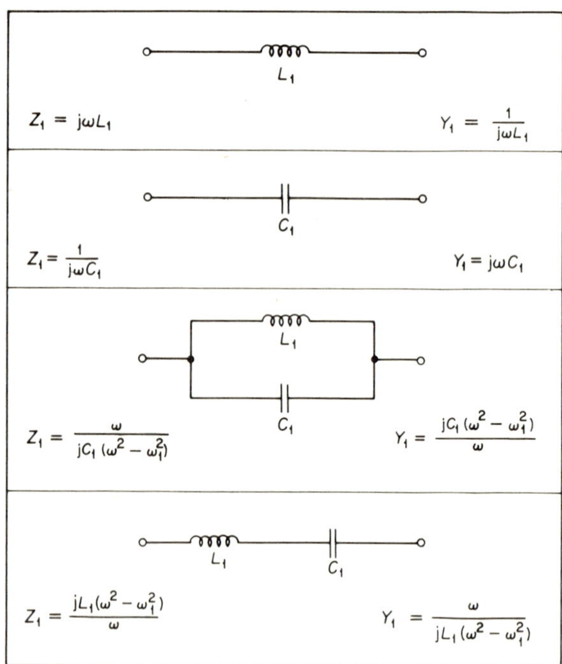

Comment

To synthesize the second Foster form, express the result as the sum of two admittances Y_1 and Y_2 and identify the components from the admittance column in Table 1.

EXAMPLE 3

Design a series-type Foster network to have an impedance j500 at $\omega = 0.5 \times 10^6$ rad/s. There is to be a pole at $\omega = 1.0 \times 10^6$ rad/s and a zero at $\omega = 2.0 \times 10^6$ rad/s.

Solution

Let

$$\omega_1 = 1.0 \times 10^6 \text{ rad/s}$$

$$\omega_2 = 2.0 \times 10^6 \text{ rad/s}$$

Since the impedance has a positive value at $\omega = 0.5 \times 10^6$ rad/s, it must decrease to zero at $\omega = 0$. Hence, there is an *external* zero at $\omega = 0$ and the type can be identified as Type 2 of Section 2.5(a).

Hence

$$Z = j\omega H \left[\frac{(\omega^2 - \omega_2^2)}{(\omega^2 - \omega_1^2)} \right]$$

or

$$j500 = j0.5 \times 10^6 \times H \left[\frac{(0.5 \times 10^6)^2 - (2 \times 10^6)^2}{(0.5 \times 10^6)^2 - (1 \times 10^6)^2} \right]$$

$$= j0.5 \times 10^6 \times H \left[\frac{(0.25) - 4}{(0.25) - 1} \right]$$

$$= j0.5 \times 10^6 H \times \tfrac{15}{3}$$

or

$$H = 2 \times 10^{-4}$$

Now let

$$\frac{\omega^2 - \omega_2^2}{\omega^2 - \omega_1^2} = 1 + \frac{A}{\omega^2 - \omega_1^2}$$

or

$$\omega^2 - \omega_2^2 = \omega^2 - \omega_1^2 + A$$

If $\omega = \omega_1$ then

$$A = \omega_1^2 - \omega_2^2 = (1.0 \times 10^6)^2 - (2 \times 10^6)^2 = -3 \times 10^{12}$$

Hence

$$Z = j\omega H \left[1 + \frac{A}{\omega^2 - \omega_1^2} \right]$$

$$Z = j\omega \times 2 \times 10^{-4} - \frac{j\omega \times 6 \times 10^8}{(\omega^2 - \omega_1^2)}$$

or

$$= j\omega \times 2 \times 10^{-4} + \frac{\omega \times 6 \times 10^8}{j(\omega^2 - \omega^2)}$$

From Table 1 this corresponds to an inductance L in series with a parallel L_1 and C_1.

Hence $\quad L = 2 \times 10^{-4}$

and $\quad C_1 = \dfrac{1}{6 \times 10^8} = 1700 \text{ pf}$

Now $\quad \omega_1^2 = \dfrac{1}{L_1 C_1}$

or $\quad L_1 = \dfrac{1}{(1 \times 10^6)^2 \times 1700 \times 10^{-12}} = 0{\cdot}6 \text{ mH}$

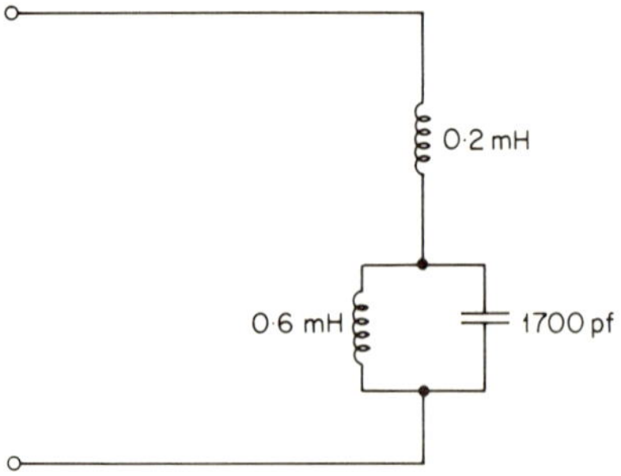

Fig. 9

EXAMPLE 4

State and explain Foster's reactance theorem. Sketch the variation of reactance with frequency for the loss-free one-port network in Fig. 10.

Express the reactance in the form

$$X = \frac{k}{\omega} \frac{(\omega^2 - \omega_1^2)(\omega^2 - \omega_3^2)}{(\omega^2 - \omega_2^2)}$$

and derive expressions for k, ω_1, ω_2 and ω_3 in terms of L_1, C_1, L_3 and C_3.

L.U.B.Sc(Eng) Tels. 1968

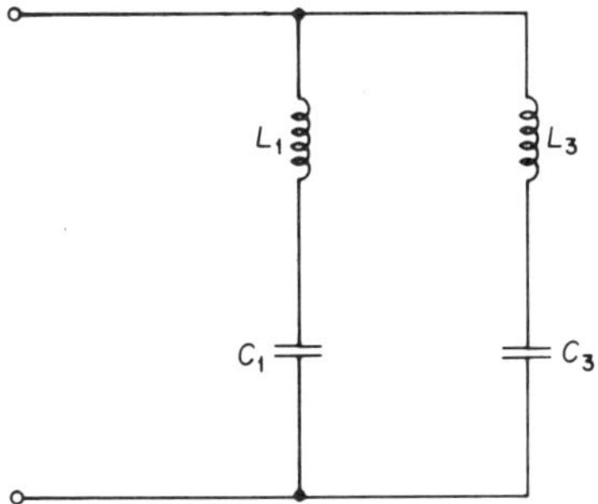

Fig. 10

Solution
The answer to the first part of the question will be found in Section 2.5.

The expression for X shows that the network has zeros at ω_1 and ω_3 and a pole at ω_2. Also, the input impedance (or reactance) tends to a large value at $\omega = 0$ and $\omega = \infty$ and so these are the external poles. Hence, the reactance diagram can be drawn as in Fig. 11.

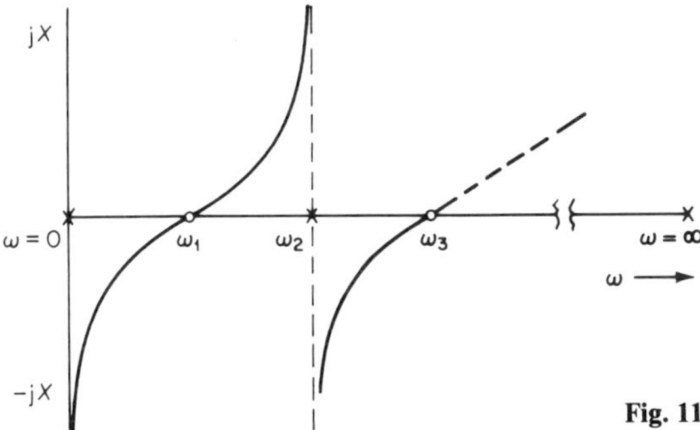

Fig. 11

The input impedance of the network is by inspection

$$Z = \frac{Z_1 \cdot Z_2}{Z_1 + Z_2}$$

with
$$Z_1 = j[\omega L_1 - (1/\omega)C_1]$$

$$Z_2 = j[\omega L_3 - (1/\omega)C_3]$$

Hence
$$jX = \frac{j[\omega L_1 - (1/\omega)C_1] \cdot j[\omega L_3 - (1/\omega)C_3]}{j[\omega(L_1 + L_3) - 1/\omega(1/C_1 + 1/C_3)]}$$

or
$$X = \frac{1}{\omega} \cdot \frac{[(\omega^2 L_1 C_1 - 1)(\omega^2 L_3 C_3 - 1)]}{[\omega^2(L_1 + L_3)C_1 C_3 - (C_1 + C_3)]}$$

$$= \frac{1}{\omega} \cdot \frac{L_1 C_1 \cdot L_3 C_3[(\omega^2 - 1/L_1 C_1)(\omega^2 - 1/L_3 C_3)]}{(L_1 + L_3)[\omega^2 - (C_1 + C_3)/C_1 C_3(L_1 + L_3)]}$$

$$= \frac{L_1 L_3}{\omega(L_1 + L_3)} \left[\frac{[\omega^2 - (1/L_1 C_1)][\omega^2 - (1/L_3 C_3)]}{\omega^2 - (C_1 + C_3)/C_1 C_3(L_1 + L_3)} \right]$$

Equating this to the expression given we have

$$\frac{k}{\omega} \frac{(\omega^2 - \omega_1^2)(\omega^2 - \omega_3^2)}{(\omega^2 - \omega_2^2)}$$

$$\equiv \frac{L_1 L_3}{\omega(L_1 + L_3)} \left[\frac{[\omega^2 - (1/L_1 C_1)][\omega^2 - (1/L_3 C_3)]}{\omega^2 - (C_1 + C_3)/C_1 C_3(L_1 + L_3)} \right]$$

Hence

$$k = L_1 L_3/(L_1 + L_3)$$

$$\omega_1 = \frac{1}{\sqrt{L_1 C_1}}$$

$$\omega_2 = \sqrt{(C_1 + C_3)/C_1 C_3(L_1 + L_3)} = \sqrt{\frac{1}{(L_1 + L_3)C_1 C_3/(C_1 + C_3)}}$$

$$\omega_3 = \frac{1}{\sqrt{L_3 C_3}}$$

(c) Synthesis of Cauer networks[3–4]

This method uses the ladder-type network and the continuous fraction expansion, which yields two canonic forms. As an example, consider the general ladder network shown in Fig. 12 with series arms expressed as impedances and shunt arms as admittances.

Here
$$Z(s) = Z_1(s) + \frac{1}{Y_2(s) + Y_3(s)}$$

$$= Z_1(s) + \cfrac{1}{Y_2(s) + \cfrac{1}{Z_3(s) + \cfrac{1}{Y_4(s) + Y_5(s)}}}$$

The expression on the right-hand side is called a continuous fraction expansion in which every other term is either an impedance or an admittance of the given ladder network to be synthesized.

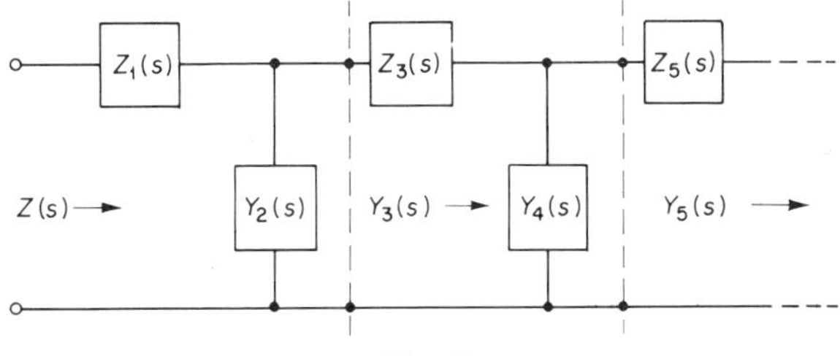

Fig. 12

Hence, by writing the given expression for Z in terms of an expansion like this, we can identify each of the components of the network.

Since $Z_1(s)$ is an inductance, it contains a term in s only and has a pole at $s = \infty$. By writing $Z_1(s)$ separately, it amounts to 'removing a pole' from Z and obtaining the remainder $Z - Z_1(s) = 1/[Y_2(s) + Y_3(s)]$.

$Y_3(s)$ is then inverted and another pole $Z_3(s)$ is removed to leave the remainder $1/[Y_4(s) + Y_5(s)]$ and so on.

This technique yields one of the canonic forms and by removing 'zeros' rather than 'poles', an alternative canonic form can be obtained. It is best illustrated in the following example.

EXAMPLE 5

Synthesize the two types of Cauer networks which have the input impedance $Z(s)$ given by

$$Z(s) = \frac{3s^4 + 45s^2 + 50}{s^3 + 10s}$$

where $s = j\omega$.

First form

Carrying out the division starting with the *highest* power in s first, we obtain

$$s^3 + 10s)\overline{3s^4 + 45s^2 + 50}(3s \to Z_1(s)$$
$$\underline{3s^4 + 30s^2}$$
$$15s^2 + 50)\overline{s^3 + 10s}(s/15 \to Y_2(s)$$
$$\underline{s^3 + \tfrac{10}{3}s}$$
$$\tfrac{20}{3}s)\overline{15s^2 + 50}(9s/4 \to Z_3(s)$$
$$\underline{15s^2}$$
$$50)\tfrac{20}{3}s(\tfrac{2}{15}s \to Y_4(s)$$
$$\underline{\tfrac{20}{3}s}$$

Second form
Carrying out the division starting with the *lowest* power in s first, we obtain

$$10s + s^3 \overline{)50 + 45s^2 + 3s^4}(5/s \rightarrow Z_1(s)$$
$$\underline{50 + 5s^2}$$
$$40s^2 + 3s^4 \overline{)10s + s^3}(\tfrac{1}{4}s \rightarrow Y_2(s)$$
$$\underline{10s + \tfrac{3}{4}s^3}$$
$$s^3/4 \overline{)40s^2 + 3s^4}(160/s \rightarrow Z_3(s)$$
$$\underline{40s^2}$$
$$3s^4 \overline{)s^3/4}(\tfrac{1}{12}s \rightarrow Y_4(s)$$
$$\underline{s^3/4}$$

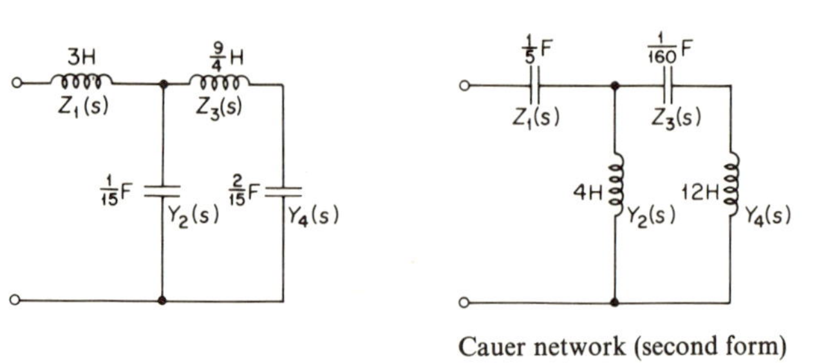

Cauer network (second form)

Fig. 13

Comments
(i) Arranging $Z(s)$ in *descending* powers of s (first form) ensures that the first term of the quotient will be a term in s to give a pole at $s = \infty$, i.e. *poles* at infinity are being removed by the division.
(ii) Arranging $Z(s)$ in *ascending* powers of s (second form) ensures that the first term of the quotient will be a term in $1/s$ to give a zero at $s = \infty$, i.e. *zeros* at infinity are being removed by the division.

3
Two-port networks

3.1. Symmetrical networks

A symmetrical two-port network has an input port and an output port. The input and output ports are similar and may be interchanged. Its two most important properties are the characteristic impedance Z_0 and propagation coefficient γ.

(a) Characteristic impedance

This is defined as the input impedance of an infinite number of two-port networks in cascade or of one network terminated in Z_0. It is illustrated in Fig. 14.

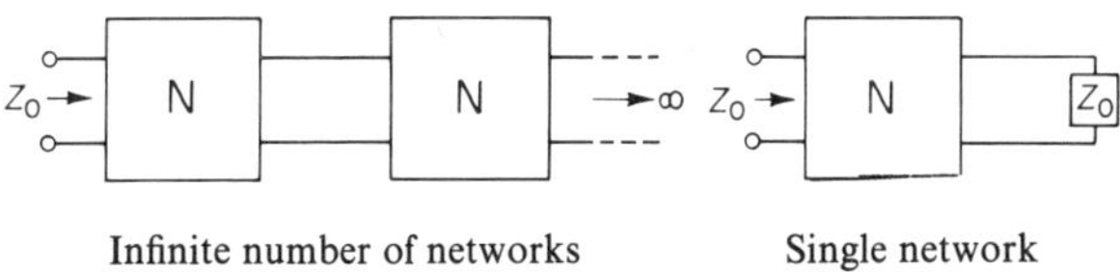

Fig. 14

(b) Propagation coefficient

It is also known as the iterative transfer coefficient and defined as the natural logarithm of the ratio of input and output currents of a network terminated on an iterative basis at both ends, i.e. in Z_0.

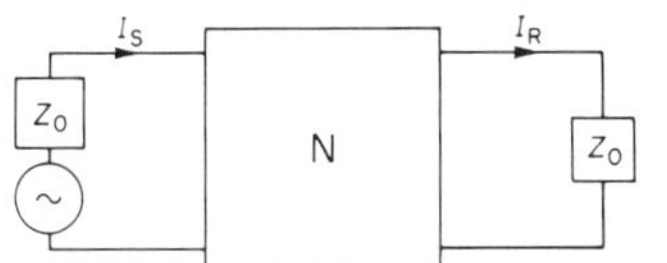

Fig. 15

Consider the network of Fig. 15 with an input current I_S and an output current I_R. In general, I_S will be greater than I_R and will have an angle different from I_S. The ratio I_S/I_R is therefore a complex vector quantity and can be written as e^γ where γ is a complex quantity.

Hence, let $\qquad I_S/I_R = e^\gamma = e^{\alpha + j\beta} = e^\alpha \cdot e^{j\beta} = e^\alpha\underline{/\beta}$

where $\qquad\qquad\qquad \gamma = \alpha + j\beta$

We have $\qquad\qquad\qquad \left|\dfrac{I_S}{I_R}\right| = e^\alpha$

or $\qquad\qquad\qquad \alpha = \log_e \left|\dfrac{I_S}{I_R}\right|$ nepers

Now 1 neper $= 8.686$ db, giving

$$\alpha = 2.3 \times 8.686 \log_{10} \left|\frac{I_S}{I_R}\right| \text{ db}$$

or $\qquad\qquad\qquad \alpha = 20 \log_{10} \left|\dfrac{I_S}{I_R}\right| \text{ db}$

Since α is responsible for the attenuation of the network it is called the *attenuation coefficient*. The phase shift through the network is due to β which is called the *phase-change coefficient* and it is measured in radians or degrees.

Comments

For n similar networks in cascade we have:

(i) Total attenuation $= \alpha + \alpha + \cdots = n\alpha$ nepers or decibels.

(ii) Total phase shift $= \beta + \beta + \cdots = n\beta$ radians or degrees.

(c) Insertion loss

When the network of Fig. 16 is inserted between a generator and load, there is a loss of power in the load due to the mismatch between the generator and network, the network and load and the attenuation through the network. The total loss is called the insertion loss. Hence the insertion loss is the sum of the mismatch losses and attenuation loss.

If P_1 is the power in the load without the network inserted and P_2 is the power in the load with the network inserted, the insertion loss is defined as

$$\text{insertion loss} = 10 \log \frac{P_1}{P_2} \text{ db}$$

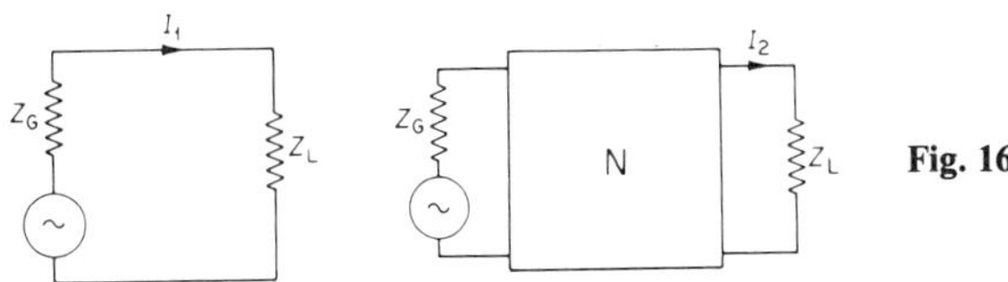

Fig. 16

Here
$$P_1 = I_1^2 \cdot Z_L \text{ (real part)}$$
$$P_2 = I_2^2 \cdot Z_L \text{ (real part)}$$

Hence,
$$\text{insertion loss} = 10 \log \frac{|I_1^2|}{|I_2^2|} = 20 \log_{10} \frac{|I_1|}{|I_2|} \text{ db}$$

where I_1, I_2 are the load currents *without* the network inserted and *with* the network inserted, respectively.

3.2 Typical networks

(a) General ladder network

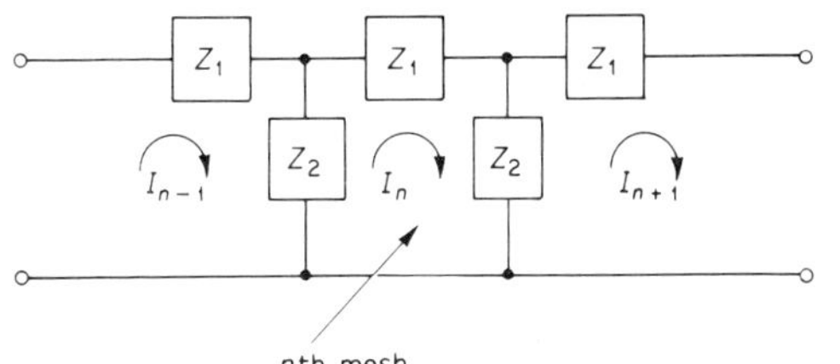

Fig. 17

For the configuration shown in Fig. 17, we have for the nth mesh

$$(-I_{n-1}) \cdot Z_2 + I_n Z_2 + I_n Z_1 + I_n Z_2 - (I_{n+1}) \cdot Z_2 = 0$$

Dividing through by $-2Z_2 I_n$ gives

$$\left\{ \frac{I_{n-1}}{2I_n} \right\} - \frac{(Z_1 + 2Z_2)}{2Z_2} + \frac{I_{n+1}}{2I_n} = 0$$

By definition,

$$\frac{I_{n-1}}{I_n} = e^\gamma \quad \text{and} \quad \frac{I_n}{I_{n+1}} = e^\gamma$$

Hence
$$\frac{e^{\gamma}}{2} + \frac{e^{-\gamma}}{2} = 1 + \frac{Z_1}{2Z_2}$$

or
$$\cosh \gamma = 1 + \frac{Z_1}{2Z_2}$$

and
$$\gamma = \cosh^{-1}\left(1 + \frac{Z_1}{2Z_2}\right)$$

from which γ can be evaluated.

Comment
This expression is also valid for a T or π network, since a ladder network can be subdivided into a series of T or π networks.

(b) T network

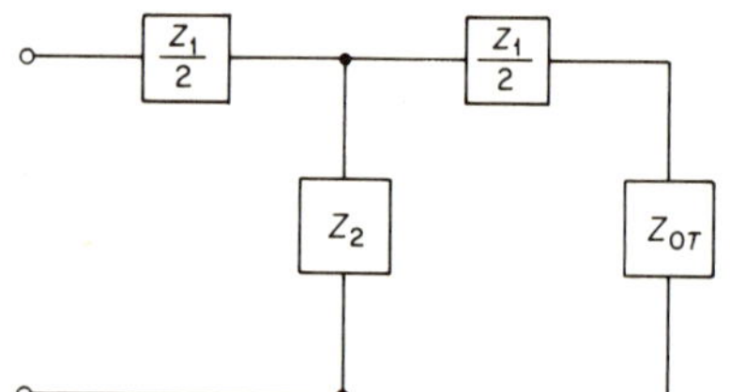

Fig. 18

Let Z_{OT} be the characteristic impedance of the network in Fig. 18 which is correctly terminated.

We have
$$Z_{1n} = \frac{Z_1}{2} + \frac{Z_2\left(\dfrac{Z_1}{2} + Z_{OT}\right)}{Z_2 + \dfrac{Z_1}{2} + Z_{OT}}$$

which equals Z_{OT} by definition.

   Hence, expanding the right-hand side and equating to Z_{OT} gives

$$\frac{Z_1Z_2}{2} + \frac{Z_1^2}{4} + \frac{Z_1Z_{OT}}{2} + \frac{Z_1Z_2}{2} + Z_2Z_{OT} = Z_2Z_{OT} + \frac{Z_1Z_{OT}}{2} + Z_{OT}^2$$

or
$$Z_1^2/4 + Z_1Z_2 = Z_{OT}^2$$

or
$$Z_{OT} = \sqrt{Z_1Z_2 + Z_1^2/4}$$

Now let Z_{oc} and Z_{sc} be the open-circuit and short-circuit impedances of the network.

Hence
$$Z_{oc} = \frac{Z_1}{2} + Z_2$$

$$Z_{sc} = \frac{Z_1}{2} + \frac{Z_2 \cdot Z_1/2}{Z_2 + Z_1/2} = \frac{Z_1 Z_2 + Z_1^2/4}{Z_2 + Z_1/2}$$

Hence
$$Z_{oc} \cdot Z_{sc} = Z_1 \cdot Z_2 + Z_1^2/4$$

which equals Z_{0T}^2.

Therefore
$$Z_{0T} = \sqrt{Z_{oc} \cdot Z_{sc}}$$

Now from Section 3.2(a) we have

$$\cosh \gamma = 1 + \frac{Z_1}{2Z_2} = \frac{Z_2 + Z_1/2}{Z_2} = \frac{Z_{oc}}{Z_2} \quad \text{from above}$$

Also
$$\sinh \gamma = \sqrt{\cosh^2 \gamma - 1}$$

$$= \sqrt{\left(1 + \frac{Z_1}{2Z_2}\right)^2 - 1}$$

$$= \sqrt{\frac{Z_1 Z_2 + Z_1^2/4}{Z_2^2}} = \frac{Z_{0T}}{Z_2}$$

Hence
$$\tanh \gamma = \frac{\sinh \gamma}{\cosh \gamma} = \frac{Z_{0T}}{Z_2} \cdot \frac{Z_2}{Z_{oc}} = \frac{Z_{0T}}{Z_{oc}} = \frac{\sqrt{Z_{oc} \cdot Z_{sc}}}{Z_{oc}}$$

or
$$\tanh \gamma = \sqrt{Z_{sc}/Z_{oc}}$$

Comment

To evaluate γ which is complex the following method is used.

Since $\gamma = \alpha + j\beta$, α and β being real quantities are evaluated from well-known trigonometric relationships given below.

Let $\tanh \gamma = \tanh(\alpha + j\beta) = A + jB$ since it is complex.

Hence,
$$\tanh 2\alpha = \frac{2A}{1 + A^2 + B^2}$$

$$\tan 2\beta = \frac{2B}{1 - A^2 - B^2}$$

from which α and β are obtained using standard tables.

(c) π **Network**

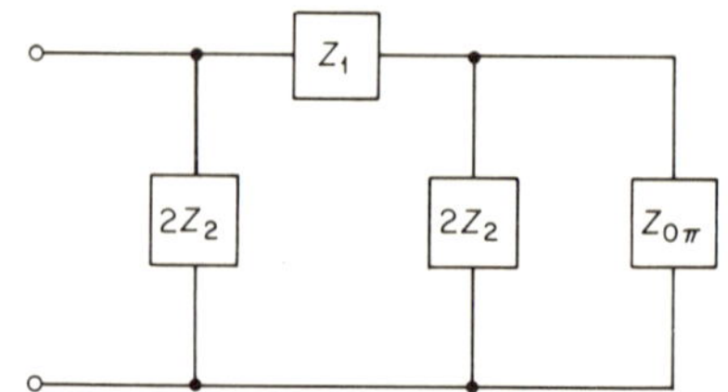

Fig. 19

Let $Z_{0\pi}$ be the characteristic impedance of the network which is correctly terminated.

Hence
$$Z_{in} = \frac{2Z_2\left[Z_1 + \dfrac{2Z_2 \cdot Z_{0\pi}}{2Z_2 + Z_{0\pi}}\right]}{2Z_2 + Z_1 + \dfrac{2Z_2 \cdot Z_{0\pi}}{2Z_2 + Z_{0\pi}}}$$

which equals $Z_{0\pi}$ by definition.

Expanding the right-hand side and equating to $Z_{0\pi}$ gives

$$4Z_2^2 Z_1 + 2Z_2 Z_1 Z_{0\pi} + 4Z_2^2 \cdot Z_{0\pi} = 4Z_2^2 Z_{0\pi} + 2Z_2 Z_{0\pi}^2 + 2Z_1 Z_2 Z_{0\pi}$$
$$+ Z_1 Z_{0\pi}^2 + 2Z_2 Z_{0\pi}^2$$

or
$$\frac{4Z_1 Z_2^2}{4Z_2 + Z_1} = Z_{0\pi}^2$$

Hence
$$Z_{0\pi} = \frac{Z_1 Z_2}{\sqrt{(Z_1 Z_2 + Z_1^2)/4}} = \frac{Z_1 Z_2}{Z_{0T}}$$

Now let Z_{oc} and Z_{sc} be the open-circuit and short-circuit impedances respectively.

Here
$$Z_{oc} = \frac{2Z_2(Z_1 + 2Z_2)}{2Z_2 + Z_1 + 2Z_2}$$

$$= \frac{2Z_2(Z_1 + 2Z_2)}{Z_1 + 4Z_2}$$

and
$$Z_{sc} = \frac{2Z_2 \cdot Z_1}{(2Z_2 + Z_1)}$$

Therefore $Z_{oc} \cdot Z_{sc} = \dfrac{2Z_2 \cdot 2Z_2 Z_1}{Z_1 + 4Z_2} = \dfrac{Z_1^2 \cdot Z_2^2}{Z_1 Z_2 + (Z_1^2/4)} = Z_{0\pi}^2$

or

$$Z_{0\pi} = \sqrt{Z_{oc} \cdot Z_{sc}}$$

Again, from Section 3.2(a) we have

$$\cosh \gamma = 1 + \frac{Z_1}{2Z_2} = \frac{2Z_2 + Z_1}{2Z_2} = \frac{Z_1}{Z_{sc}}$$

and

$$\sinh \gamma = \sqrt{\cosh^2 \gamma - 1}$$

$$= \sqrt{\frac{Z_1}{Z_2} + \frac{Z_1^2}{4Z_2^2}}$$

$$= \frac{\sqrt{Z_1 Z_2 + (Z_1^2/4)}}{Z_2} = \frac{Z_1}{Z_{0\pi}}$$

Therefore

$$\tanh \gamma = \frac{\sinh \gamma}{\cosh \gamma}$$

$$= \frac{Z_1}{Z_{0\pi}} \times \frac{Z_{sc}}{Z_1} = \frac{Z_{sc}}{\sqrt{Z_{oc} \cdot Z_{sc}}}$$

or

$$\tanh \gamma = \sqrt{\frac{Z_{sc}}{Z_{oc}}}$$

from which γ can be evaluated as for the T network.

Comments

1. $\cosh \gamma = 1 + (Z_1/2Z_2)$ for T, π or ladder networks.

2. $\tanh \gamma = \sqrt{\dfrac{Z_{sc}}{Z_{oc}}}$ for T or π networks.

3. $Z_{0T} \cdot Z_{0\pi} = Z_1 \cdot Z_2$.

EXAMPLE 6

Deduce the relationships between the impedances of a symmetrical π network and those of the equivalent T network.

The circuit shown in Fig. 20 represents a particular π network. Determine from first principles the impedances of the equivalent T network for a frequency of 800 Hz.

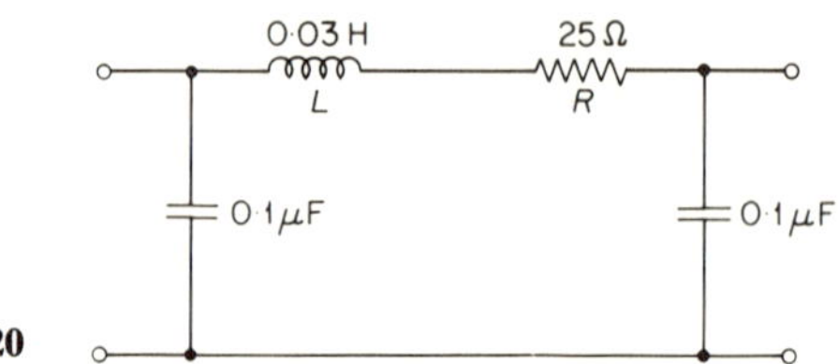

Fig. 20

If the resistance of one branch of the equivalent network is found to be negative, how can this effect be incorporated in the actual equivalent network.   L.U.B.Sc(Eng) Tels., 1962

Solution

Let the respective impedances of a T network and a π network be as shown in Fig. 21.

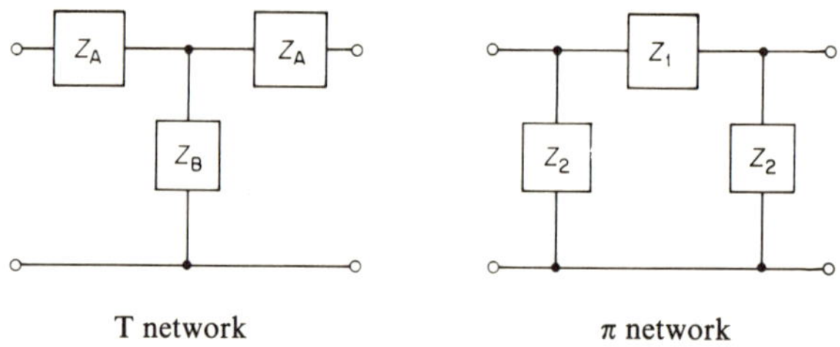

T network π network

Fig. 21

For the two networks to be *equivalent*, impedances looking into corresponding terminals must be the same. Hence, looking into terminals (1) and (2) we have

For the T network   *For the π network*

$$(Z_A + Z_B) = \frac{Z_2(Z_1 + Z_2)}{Z_1 + 2Z_2}$$

Looking into terminals (1) and (3) we have

For the T network   *For the π network*

$$(Z_A + Z_A) = \frac{Z_1 \cdot 2Z_2}{Z_1 + 2Z_2}$$

or

$$Z_A = \frac{Z_1 \cdot Z_2}{Z_1 + 2Z_2}$$

Substituting for Z_A in the equation for $(Z_A + Z_B)$ above yields

$$Z_B = \frac{Z_2^2}{Z_1 + 2Z_2}$$

Now $\qquad Z_1 = 25 + j5000 . 3 . 10^{-2} = (25 + j150)\ \Omega$

$$Z_2 = -j\frac{1}{5000 . 10^{-7}} = j2000\ \Omega$$

Hence $\qquad Z_A = \frac{(25 + j150)(-j2000)}{(25 + j150 - j4000)} = \frac{(1 + j6) . 2000}{154}\ \Omega$

$$= (13 + j78)\ \Omega$$

Hence $\qquad R_A = 13\ \Omega$

$$L_A = \frac{78}{5000} = 15.6\ \text{mH}$$

and $\qquad Z_B = \frac{(-j2000)^2}{25(1 - j154)} = \frac{-4 . 10^6(1 + j154)}{25 \times 154}$

or $\qquad R_B - j\frac{1}{\omega C_B} = (-6.75 - j1040)\ \Omega$

Hence $\qquad R_B = -6.75\ \Omega$

and $\qquad C_B = \frac{1}{5000 . 1040} = 0.192\ \mu\text{F}$

The T *network*

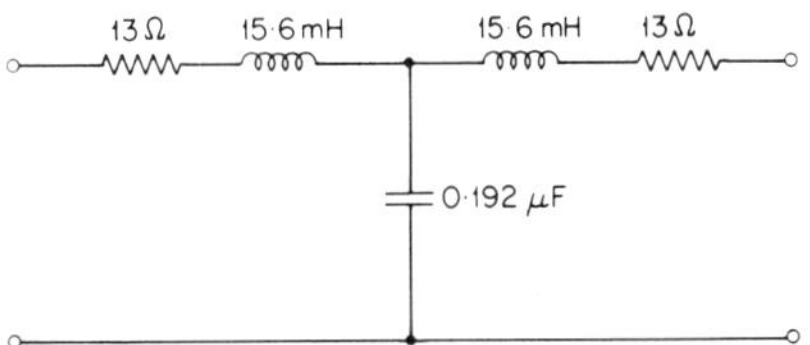

Fig. 22

Last part

If the resistance of one branch of the equivalent network is found to be negative, then this corresponds to power *generation*, since a positive resistance is associated with power *dissipation*. It can be incorporated in the branch by drawing a voltage generator in series with the resistance of that branch.

EXAMPLE 7

Define the term *insertion loss* for a transmission network. Draw the diagram of a circuit for measuring, by comparison with a calibrated attenuator, the loss of a network inserted between equal resistive terminations.

Calculate the loss in db of an *LC* network inserted between the source and load as shown in Fig. 23.   L.U.B.Sc(Eng) Tels., 1964

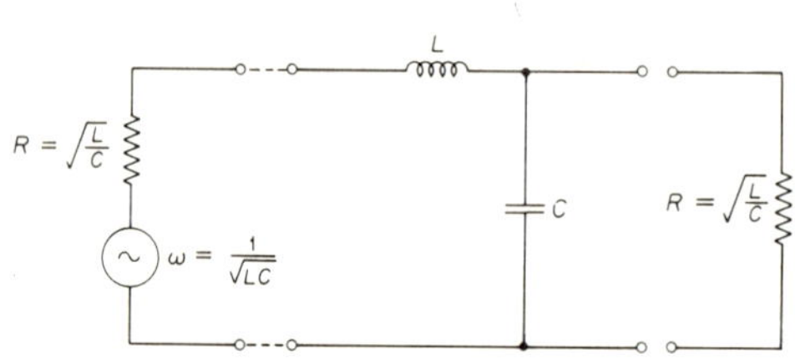

Fig. 23

When a network is inserted between a generator and load, the power loss in the load is called the insertion loss. Hence, if P_1 is the power in the load *without* the network inserted and P_2 is the power in the load *with* the network inserted, then the insertion loss is $10 \log P_1/P_2$ db.

To measure the insertion loss the circuit shown in Fig. 24 may be used, such that the network or the calibrated attenuator may be suitably switched into circuit separately, together with a power meter (or milliammeter) and the proper terminating load.

With the network inserted first, the generator voltage is adjusted to give a suitable value of power through the load or a suitable load current (if a milliammeter is used). The network is then switched out and the attenuator is switched in.

The attenuator, which is usually calibrated in db, is adjusted till the power meter (or milliammeter) reads the same value as for the previous measurement. The insertion loss is then read directly in db from the attenuator setting (see Fig. 24).

Solution

(a) Without network

Let I_1 be the load current through the load $R = \sqrt{L/C}$ and the generator voltage V.

Fig. 24

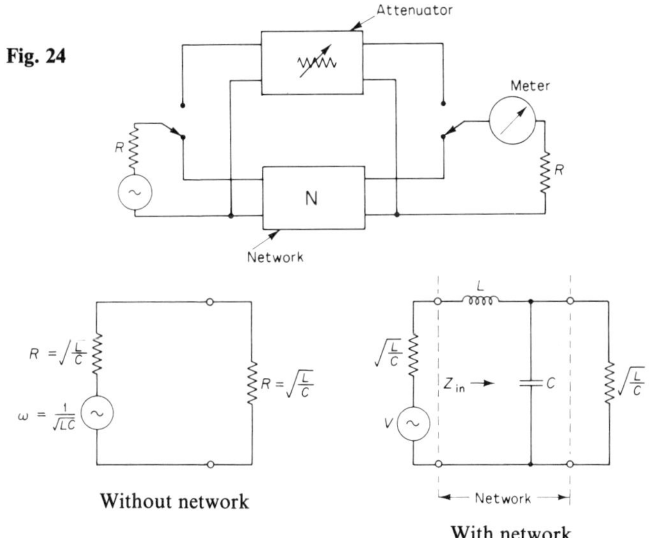

$$R = \sqrt{\tfrac{L}{C}}$$

$$\omega = \tfrac{1}{\sqrt{LC}}$$

$$R = \sqrt{\tfrac{L}{C}}$$

Without network

$$\sqrt{\tfrac{L}{C}}$$

$$Z_{in} \rightarrow \qquad C$$

$$\sqrt{\tfrac{L}{C}}$$

$$V$$

$$L$$

Network

With network

Fig. 25

Hence
$$|I_1| = \frac{V}{2R}$$

(b) With network

$$Z_{in} = j\omega L + \frac{(1/j\omega C)\,.\,R}{(1/j\omega C) + R}$$

$$= j\omega L + \frac{R}{1 + j\omega CR}$$

Hence
$$I = \frac{V}{R + Z_{in}} = \frac{V}{(R + j\omega L) + R/(1 + j\omega CR)}$$

$$= \frac{V(1 + j\omega CR)}{(R + j\omega L)(1 + j\omega CR) + R}$$

and
$$I_2 = \left[\frac{1/j\omega C}{(1/j\omega C) + R}\right]I$$

$$= \frac{1}{(1 + j\omega CR)}\left[\frac{V(1 + j\omega CR)}{(R + j\omega L)(1 + j\omega CR) + R}\right]$$

$$= \frac{V}{R + j\omega L + j\omega CR^2 - \omega^2 LCR + R}$$

Since $\omega = 1/\sqrt{LC}$ and $R = \sqrt{L/C}$ this gives

$$I_2 = \frac{V}{R + jR + jR - R + R} = \frac{V}{R(1 + 2j)}$$

and

$$|I_2| = \frac{V}{R\sqrt{5}}$$

$$\text{The insertion loss} = 20 \log_{10} \left|\frac{I_1}{I_2}\right|$$

$$= 20 \log_{10} \frac{V}{2R} \cdot \frac{R\sqrt{5}}{V}$$

$$= 10 \log_{10} \tfrac{5}{4}$$

$$= 10 \log_{10} 1 \cdot 25$$

$$\text{or insertion loss} = 0 \cdot 969 \text{ db}$$

(d)   Lattice network

Let its characteristic impedance equal Z_0 where $Z_0 = \sqrt{Z_{oc} \cdot Z_{sc}}$. From the equivalent circuit we have

$$Z_{oc} = \frac{Z_1 + Z_2}{2}$$

$$Z_{sc} = \frac{2Z_1 \cdot Z_2}{Z_1 + Z_2}$$

Hence

$$Z_0^2 = Z_1 \cdot Z_2$$

or

$$Z_0 = \sqrt{Z_1 \cdot Z_2}$$

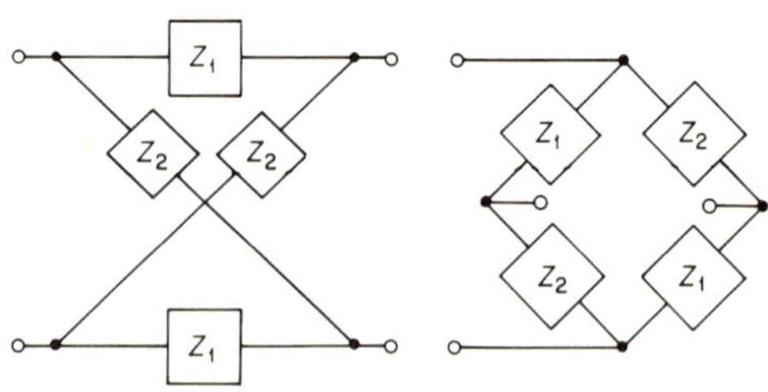

Fig. 26   Lattice network   Equivalent circuit

Now
$$\tanh \gamma = \sqrt{\frac{Z_{sc}}{Z_{oc}}}$$

$$= \frac{2\sqrt{Z_1 \cdot Z_2}}{Z_1 + Z_2}$$

$$= \frac{2Z_1 \cdot Z_0}{Z_1^2 + Z_0^2}$$

or
$$\frac{e^{2\gamma} - 1}{e^{2\gamma} + 1} = \frac{2Z_1 Z_0}{Z_1^2 + Z_0^2}$$

Hence we obtain
$$e^{2\gamma} = \frac{(Z_0 + Z_1)^2}{(Z_0 - Z_1)^2}$$

or
$$e^{\gamma} = \frac{Z_0 + Z_1}{Z_0 - Z_1} \quad \text{(using the positive value)}$$

Now
$$\tanh \frac{\gamma}{2} = \frac{e^{\gamma} - 1}{e^{\gamma} + 1}$$

Hence
$$\tanh \frac{\gamma}{2} = \frac{2Z_1}{2Z_0} = \frac{Z_1}{Z_0} = \sqrt{\frac{Z_1}{Z_2}}$$

(e)   Bridged T network

Let Z_0 be the characteristic impedance of the network where

$$Z_0 = \sqrt{Z_{oc} \cdot Z_{sc}}$$

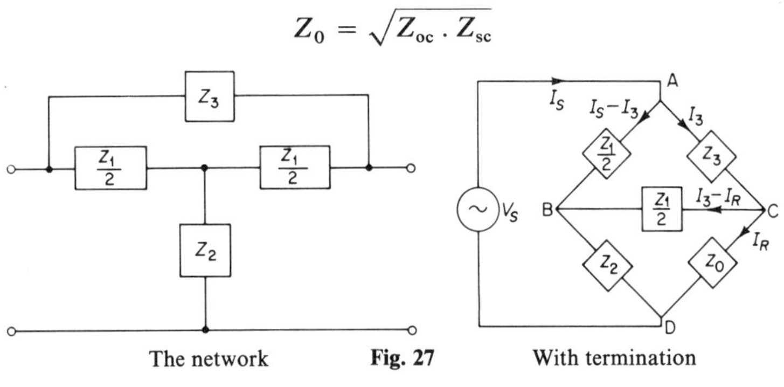

The network   **Fig. 27**   With termination

Here
$$Z_{oc} = \frac{\dfrac{Z_1}{2}\left(Z_3 + \dfrac{Z_1}{2}\right)}{\dfrac{Z_1}{2} + Z_3 + \dfrac{Z_1}{2}} + Z_2$$

$$= \frac{Z_1 Z_2 + Z_2 Z_3 + \dfrac{Z_1^2}{4} + \dfrac{Z_1 Z_3}{2}}{Z_1 + Z_3}$$

and
$$Z_{sc} = \frac{Z_3\left[\dfrac{Z_1}{2} + \dfrac{Z_2 \cdot (Z_1/2)}{Z_2 + (Z_1/2)}\right]}{Z_3 + \dfrac{Z_1}{2} + \dfrac{Z_2 \cdot (Z_1/2)}{Z_2 + (Z_1/2)}}$$

$$= \frac{Z_3 \dfrac{Z_1^2}{4} + Z_3 Z_2 Z_1}{Z_3 \dfrac{Z_1}{2} + Z_2 Z_3 + \dfrac{Z_1^2}{4} + Z_1 Z_2}$$

Hence
$$Z_0 = \sqrt{\frac{\dfrac{Z_1^2 Z_3}{4} + Z_1 Z_2 Z_3}{Z_1 + Z_3}} = \sqrt{\left(\dfrac{Z_1^2}{4} + Z_1 Z_2\right)\dfrac{Z_3}{(Z_1 + Z_3)}}$$

Usually
$$\frac{Z_1}{2} = \sqrt{Z_2 \cdot Z_3}$$

Hence
$$Z_0 = \sqrt{Z_2(Z_1 + Z_3)\frac{Z_3}{Z_1 + Z_3}}$$

or
$$Z_0 = \sqrt{Z_2 \cdot Z_3}$$

and
$$\frac{Z_1}{2} = \sqrt{Z_2 \cdot Z_3} = Z_0 \text{ also}$$

Now assume the network is terminated in Z_0 and rearranged as in Fig. 27. A voltage V_s is applied to it and for the currents shown we obtain for branches AC and CD

$$I_3 Z_3 + I_R Z_0 = V_s = I_s Z_0$$

and for branches AB, BC and CD

$$I_s \frac{Z_1}{2} - I_3 Z_1 + I_R\left(\frac{Z_1}{2} + Z_0\right) = V_s = I_s Z_0$$

Eliminating I_3 between these equations yields

$$I_R = \frac{I_s Z_0(Z_1 + Z_3) - I_s \dfrac{Z_1 Z_3}{2}}{Z_0 Z_1 + \dfrac{Z_1 Z_3}{2} + Z_3 Z_0}$$

$$= \frac{I_s\left[Z_0(Z_1 + Z_3) - \dfrac{Z_1 Z_3}{2}\right]}{Z_0(Z_1 + Z_3) + \dfrac{Z_1 Z_3}{2}}$$

or

$$\frac{I_s}{I_R} = \frac{Z_0(Z_1 + Z_3) + \dfrac{Z_1 Z_3}{2}}{Z_0(Z_1 + Z_3) - \dfrac{Z_1 Z_3}{2}}$$

Since $Z_1/2 = Z_0$ we obtain

$$\frac{I_s}{I_R} = \frac{Z_1 + 2Z_3}{Z_1} = 1 + \frac{Z_3}{Z_0}$$

or

$$\gamma = \log_e \frac{I_s}{I_R} = \log_e\left(1 + \frac{Z_3}{Z_0}\right)$$

But $Z_0 = \sqrt{Z_2 Z_3}$ and so alternatively

$$\gamma = \log_e\left(1 + \frac{Z_0}{Z_2}\right)$$

from which γ may be evaluated.

3.3 Asymmetrical networks

These are networks whose input and output ports are dissimilar and hence cannot be interchanged.

(a)   Iterative impedances

The input impedances at the two ends of such a network are different and called the iterative impedances. This is illustrated in Fig. 28, where it is assumed that Z_1 and Z_2 are the two iterative impedances.

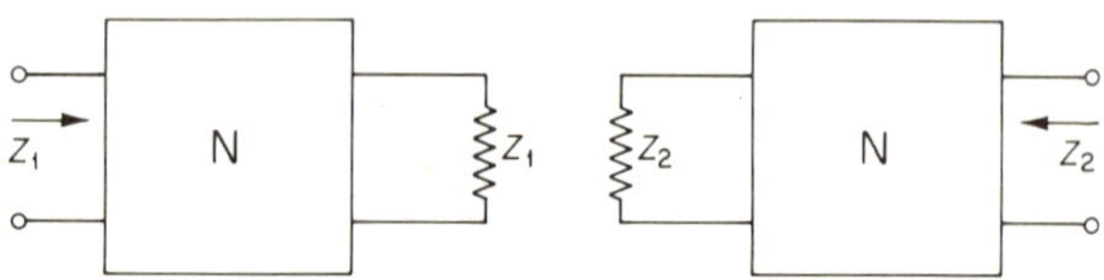

Fig. 28

(b)   Image impedances

An asymmetrical network has two different image impedances also, such that when one of them terminates the network the other is seen at the input terminals and vice versa. This is illustrated in Fig. 30, where it is assumed that Z_A and Z_B are the two image impedances.

Comments
1. Networks when used in tandem are connected on an *iterative* basis for matching purposes.
2. Networks are connected on an *image* basis when used for the maximum transfer of power between generator and load.
3. In the case of *symmetrical* networks, the two iterative impedances and the two image impedances are all equal to one another and called simply the *characteristic impedance*.

(c)   The half-section

An asymmetrical network of particular importance is the L section or half-section. It is obtained by dividing a T section or π section into two. Either half constitutes an L section as shown in Fig. 29.

Such an L section has two iterative impedances and two image impedances. The image impedances have a special significance for matching purposes as is shown subsequently (Fig. 30).

Image impedances
Let Z_A and Z_B be the image impedances as shown in Fig. 30.

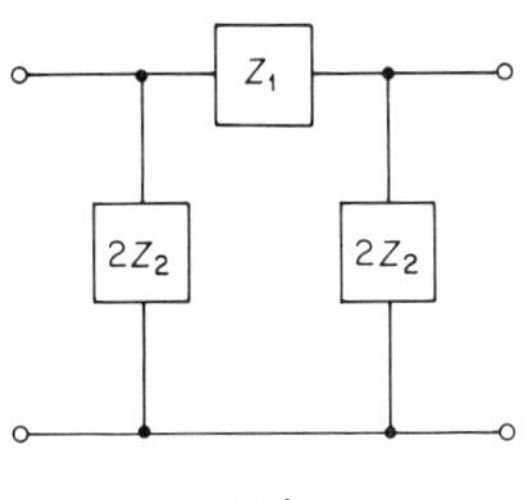

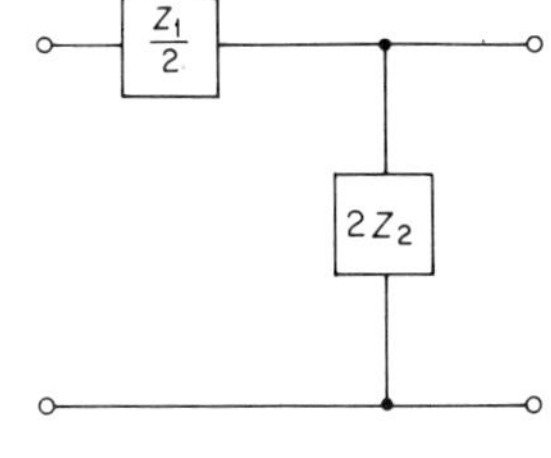

π section   **Fig. 29**   Half-section

We have
$$Z_A = \frac{Z_1}{2} + \frac{2Z_2 \cdot Z_B}{2Z_2 + Z_B}$$

Hence
$$2Z_A Z_2 + Z_A \cdot Z_B - Z_1 \cdot Z_2 - \frac{Z_1 Z_B}{2} - 2Z_2 Z_B = 0$$

or
$$Z_A Z_B + 2Z_A \cdot Z_2 - (\tfrac{1}{2}Z_1 + 2Z_2)Z_B = Z_1 Z_2$$

Again
$$Z_B = \frac{2Z_2[(Z_1/2) + Z_A]}{2Z_2 + (Z_1/2) + Z_A}$$

Hence
$$2Z_2 Z_B + \frac{Z_1 Z_B}{2} + Z_A Z_B = Z_1 Z_2 + 2Z_2 Z_A$$

or
$$Z_A Z_B - 2Z_A Z_2 + \left(\frac{Z_1}{2} + 2Z_2\right)Z_B = Z_1 Z_2$$

Subtracting this expression from the previous one for $Z_1 - Z_2$ also yields
$$4Z_A Z_2 - 2\left(\frac{Z_1}{2} + 2Z_2\right)Z_B = 0$$

or
$$\frac{Z_A}{Z_B} = \frac{(Z_1/4) + Z_2}{Z_2}$$

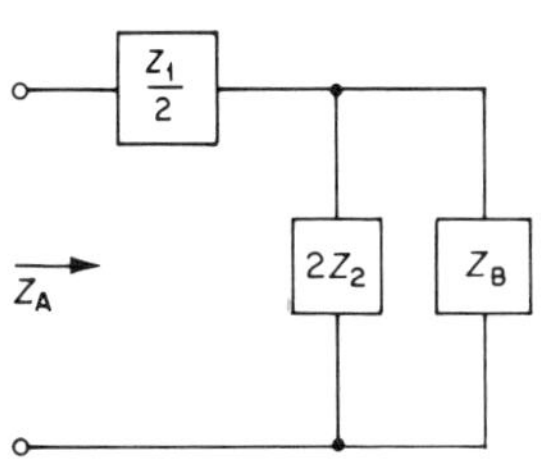

Fig. 30

Adding the two expressions for Z_1Z_2 gives

$$Z_A . Z_B = Z_1Z_2$$

Hence $$\frac{Z_A}{Z_B}(Z_A . Z_B) = \frac{(Z_1/4) + Z_2}{Z_2} . Z_1Z_2 = \frac{Z_1^2}{4} + Z_1Z_2$$

or $Z_A = \sqrt{Z_1Z_2 + (Z_1^2/4)} = Z_{0T}$ from Section 3.2(b). Similarly by division we have

$$\frac{Z_A . Z_B}{Z_A/Z_B} = \frac{Z_1Z_2 . Z_2}{[(Z_1/4) + Z_2]} = \frac{Z_1^2Z_2^2}{(Z_1^2/4) + Z_1Z_2}$$

or $Z_B = Z_1Z_2/Z_{0T} = Z_{0\pi}$ from Section 3.2(c).

Comments
1. The image impedances at the two ends of an L section are equal to the characteristic impedances of a T or π full-section respectively. Hence they can be used for matching a T network to a π network or vice versa.
2. The general formula $Z_0 = \sqrt{Z_{oc}Z_{sc}}$ applies to the two *image* impedances and can be used to determine them. It *does not* hold good for the two *iterative* impedances, which must be determined from first principles.

(d)   Image transfer coefficient, θ

This is defined as one half of the Naperian logarithm of complex ratio of the volt-amps entering and leaving the network when it is terminated in its image impedances.

$$\theta = \tfrac{1}{2}\log_e \frac{V_s . I_s}{V_R . I_R}$$

where V_s, I_s refer to the input voltage and current and V_R, I_R refer to the output voltage and current respectively.

Since θ is complex, it may likewise be expressed as $\theta = \alpha + j\beta$ where α is the image attenuation coefficient and β is the image phase-change coefficient.

EXAMPLE 8
Define the terms image impedance Z_I and image transfer coefficient γ as applied to linear passive two-terminal-pair networks.

If R_1, R_2 and R in Fig. 31 are related by $R_1 R_2 = R^2$, show that

$$Z_I = R \quad \text{and} \quad e^\gamma = 1 + \frac{R_1}{R}$$

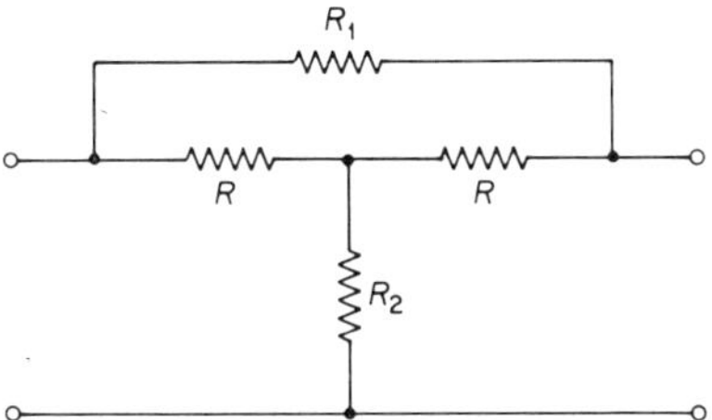

Fig. 31

The following may be used without proof:

$$Z_I^2 = Z_{oc} \cdot Z_{sc}$$

and
$$\tanh \gamma = \frac{e^{2\gamma} - 1}{e^{2\gamma} + 1} = \sqrt{\frac{Z_{sc}}{Z_{oc}}}$$

where Z_{oc} is the open-circuit driving-point impedance and Z_{sc} is the short-circuit driving-point impedance.   L.U.B.Sc(Eng) Tels., 1966

Solution
The image of an asymmetrical network are the two different impedances at the ends of the network and are such that when one impedance terminates the network the other is seen at the input terminals and vice versa.

The image transfer coefficient γ is defined as one half of the Naperian logarithm of the ratio of input volt-amps over the output volt-amps, when the network is terminated at either end in its image impedances.

Since
$$Z_I^2 = Z_{oc} \times Z_{sc}$$

Hence
$$Z_{oc} = R_2 + \frac{R(R_1 + R)}{(2R + R_1)} = \frac{2RR_2 + R_1 R + 2R^2}{2R + R_1} \qquad (\because R_1 R_2 = R^2)$$

and
$$Z_{sc} = \frac{[R + (RR_2/R + R_2)]R_1}{[R + (RR_2/R + R_2)] + R_1} = \frac{R^2 R_1 + 2RR_1 R_2}{R^2 + 2RR_2 + RR_1 + R_1 R_2}$$

$$= \frac{R_1 R^2 + 2R^3}{2R^2 + 2RR_2 + RR_1}$$

Hence

$$Z_I^2 = Z_{oc} \times Z_{sc} = \frac{(2RR_2 + RR_1 + 2R^2)}{2R + R_1} \times \frac{R_1R + 2R^3}{(2R^2 + 2RR_2 + RR_1)}$$

$$= \frac{R^2(2R + R_1)}{2R + R_1} = R^2$$

or

$$Z_I = R$$

Again

$$\tanh \gamma = \frac{e^{2\gamma} - 1}{e^{2\gamma} + 1}$$

$$= \sqrt{\frac{R_1R^2 + 2R^3}{(2R^2 + 2RR_2 + RR_1)} \times \frac{2R + R_1}{(2RR_2 + RR_1 + 2R^2)}}$$

$$= \frac{(2R + R_1)}{(2R + R_1) + 2R_2}$$

Hence we obtain

$$\frac{2e^{2\gamma}}{-2} = \frac{2(2R + R_1) + 2R_2}{-2R_2}$$

or

$$e^{2\gamma} = \frac{(2R + R_1) + R_2}{R_2}$$

$$= 1 + \frac{2R + R_1}{R_2}$$

Now

$$R_1R_2 = R^2$$

$$e^{2\gamma} = 1 + \frac{2R + R_1}{R^2/R_1}$$

$$= 1 + \frac{2RR_1 + R_1^2}{R^2}$$

$$= \left(\frac{R + R_1}{R}\right)^2$$

or

$$e^{\gamma} = \frac{R + R_1}{R}$$

$$= 1 + \frac{R_1}{R}$$

EXAMPLE 9
Explain what is meant by the term asymmetrical when applied to two-port networks. An unbalanced two-port resistive network has input terminals 1 and 2 and output terminals 3 and 4. The values of resistance measured at 1–2 when terminals 3–4 are first short-circuited and then open-circuited are respectively 275 Ω and 500 Ω. The resistance measured at 3–4 with 1–2 open-circuited is 400 Ω. Determine the equivalent T network and the image impedances. Hence calculate the insertion loss produced by the network when inserted between its image impedances.

L.U.B.Sc(Eng) Tels., 1968

Solution
An asymmetrical network is one whose input and output ports are dissimilar and therefore are not interchangeable.

Let the network consist of resistors $R_1 R_2$ and R_3 as shown in Fig. 32. From the data given we have

$$R_1 + R_2 = 500 \tag{1}$$

$$R_1 + \frac{R_2 . R_3}{R_2 + R_3} = 275 \tag{2}$$

From (2) and (3)

$$R_2 + R_3 = 400 \tag{3}$$

$$R_1 + \frac{R_2 . R_3}{400} = 275 \tag{4}$$

From (1)

$$R_2 = 500 - R_1 \tag{5}$$

From (1) and (3)

$$R_1 - R_3 = 100 \tag{6}$$

From (4), (5) and (6)

$$R_1 + \frac{(500 - R_1)(R_1 - 100)}{400} = 275$$

or $\qquad 400R_1 + 500R_1 - R_1^2 - 50{,}000 + 100R_1 = 110{,}000$

Hence $\qquad R_1^2 - 1000R_1 + 160{,}000 = 0$

or $\qquad (R_1 - 800)(R_1 - 200) = 0$

Therefore $\qquad R_1 = 800\,\Omega \quad \text{or} \quad 200\,\Omega$

$$R_2 = (500 - 800)\,\Omega \quad \text{or} \quad (500 - 200)\,\Omega$$

Hence $\qquad R_2 = 300\ \Omega$

and $\qquad R_1 = 200\ \Omega \qquad (R_1 = 800\ \Omega$ inadmissible$)$

$\qquad R_3 = 400 - R_2 = 100\ \Omega$

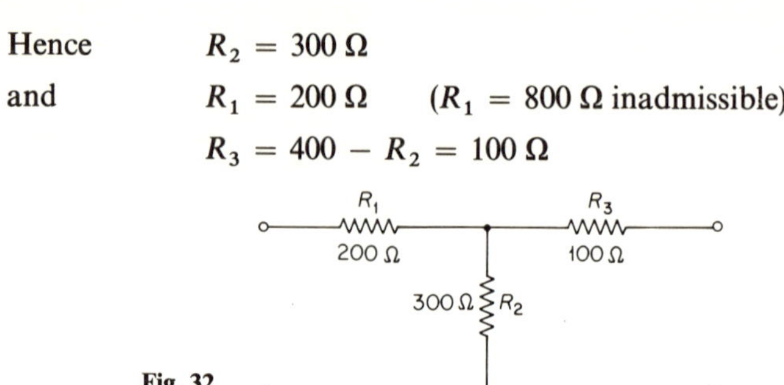

Fig. 32

The image impedances Z_A and Z_B are obtained from the expression $Z_0 = \sqrt{Z_{oc}\, .\, Z_{sc}}$ applied at the respective ends.

For Z_A $\qquad Z_{oc} = 500\ \Omega$

$$Z_{sc} = 200 + \frac{300\, .\, 100}{400} = 275\ \Omega$$

or $\qquad Z_A = \sqrt{500\, .\, 275} = 371\ \Omega$

For Z_B $\qquad Z_{oc} = 400\ \Omega$

$$Z_{sc} = 100 + \frac{300\, .\, 200}{500} = 220\ \Omega$$

$$Z_B = \sqrt{400\, .\, 220} = 297\ \Omega$$

Insertion loss.   Since the terminating impedances are the image impedances, the input generator impedance is 371 Ω and the terminating or load impedance is 297 Ω.

Without the network.   Let the generator voltage be V and the current through the load I_1.

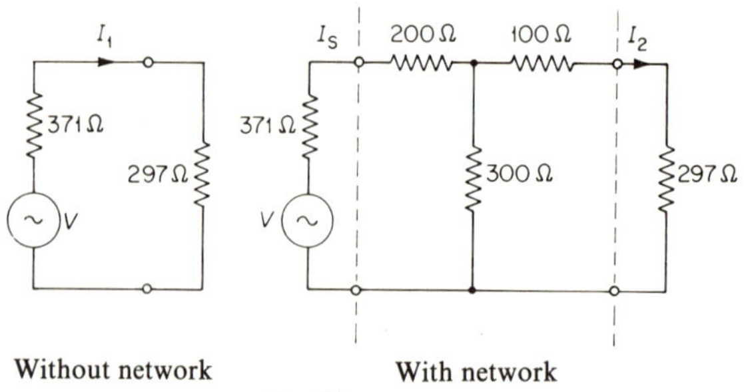

Without network   With network

Fig. 33

Hence
$$I_1 = \frac{V}{371 + 297} = \frac{V}{668}$$

With the network. Let the input current to the network be I_s and the load current be I_2 for a generator voltage V.

Hence
$$I_s = \frac{V}{371 + 371} = \frac{V}{742}$$

Also
$$I_2 = \left[\frac{300}{300 + 100 + 297}\right] I_s = \frac{300V}{697 \times 742}$$

$$\text{The insertion loss} = 20 \log \frac{I_1}{I_2}$$

$$= 20 \log \frac{V}{668} \times \frac{697 \times 742}{300V}$$

$$= 20 \log(2 \cdot 58)$$

$$= 8 \cdot 232 \text{ db}$$

4
Prototype filters

A filter is a special network whose function is to allow signals in certain bands of frequencies to pass easily, while highly attenuating or suppressing adjacent unwanted bands. The band of frequencies passed is called the *pass band* and the band of frequencies attenuated is the *stop band*.

Basic filter structures from which other more sophisticated structures can be constructed are called *prototype* filters.

4.1   Types of filter

The most common types which find wide use in communication circuits are the low-pass, high-pass, band-pass and band-stop filters.* Such filters are usually made up of T or π sections, although some, such as crystal filters, are composed of lattice sections.

4.2   Classical filter theory[5-6]

In this theory, the procedure is to take a filter structure, analyse it, and obtain as much information as possible about its characteristics. In order to do this easily, ideal components are assumed, i.e. they are considered to be pure reactances. This leads to a great simplification in the theory, which would otherwise be cumbersome. Moreover, the transition to practical components with resistance can easily be made by slight modification of the basic results obtained.

In this approach, the analysis can be qualitative and then involves simple graphical considerations, or it may be quantitative and then requires a more rigorous mathematical treatment.

The starting point is usually a consideration of the input impedance of a correctly terminated filter of T or π configuration. For pure reactances jX_1 and jX_2 of opposite sign (the signs are included in X_1 or X_2), we have

$$Z_{0T} = \sqrt{jX_1 \cdot jX_2 - \frac{X_1^2}{4}} = j\sqrt{X_1\left(\frac{X_1}{4} + X_2\right)}$$

* See Appendices A and B for descriptions of Band-pass and Band-stop filters.

and
$$Z_{0\pi} = \frac{Z_1 Z_2}{Z_{0T}} = \frac{jX_1 \cdot jX_2}{j\sqrt{X_1[(X_1/4) + X_2]}} = j^2 \frac{X_1 X_2}{Z_{0T}}$$

$$= -\frac{X_1 X_2}{Z_{0T}} = \frac{\text{a real constant}}{Z_{0T}}$$

and so the sign of $Z_{0\pi}$ depends on that of Z_{0T}.

In the case of a T filter terminated in Z_{0T} this means that power can only be dissipated in Z_{0T} if Z_{0T} is real and resistive, i.e. certain frequencies will pass through the filter and this corresponds to the *pass band*. This result can be summarised in the following rules:

RULES

1. If X_1 and $[(X_1/4) + X_2]$ are of opposite sign, then Z_{0T} is resistive and it corresponds to the *pass band*.

2. If X_1 and $[(X_1/4) + X_2]$ are of same sign, then Z_{0T} is reactive and it corresponds to the *stop band*.

(a)   Low-pass filter

It passes all frequencies from zero to a certain cut-off frequency f_c. The configuration and symbolism is shown in Fig. 34.

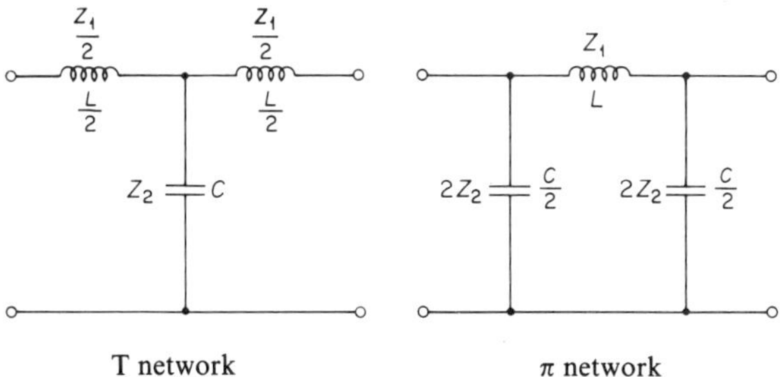

Fig. 34

For the T section we have

$$Z_1 = j\omega L \qquad \text{or} \quad X_1 = \omega L$$

$$Z_2 = -j\frac{1}{\omega C} \quad \text{or} \quad X_2 = -\frac{1}{\omega C}$$

and
$$\frac{X_1}{4} + X_2 = \frac{\omega L}{4} - \frac{1}{\omega C}$$

A plot of $[(X_1/4) + X_2]$ is shown in Fig. 35 and the cut-off frequency f_c occurs when Z_{0T} changes between reactive and resistive. Cut-off occurs when $Z_{0T} = 0$ and the filter behaves as a short-circuit.

We then have

$$\frac{\omega_c L}{4} - \frac{1}{\omega_c C} = 0$$

or

$$\omega_c^2 = \frac{4}{LC}$$

and

$$f_c = \frac{1}{\pi\sqrt{LC}}$$

the cut-off frequency for a T or π network.

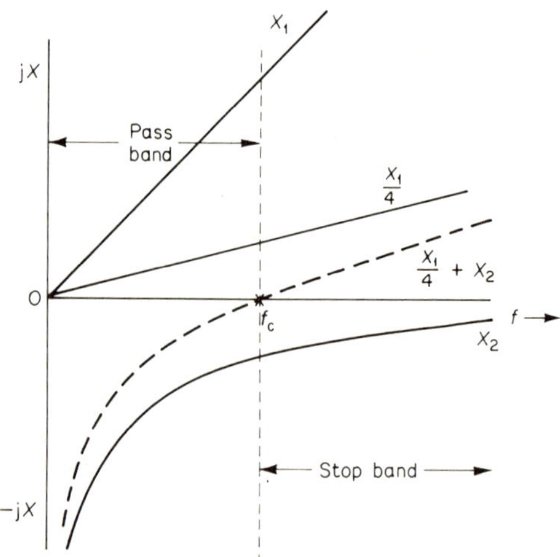

Fig. 35

In Section 3.2(a) we obtained

$$\cosh \gamma = 1 + \frac{Z_1}{2Z_2} \quad \text{for a T or } \pi \text{ network}$$

Here
$$Z_1 = j\omega L$$

$$Z_2 = -j\frac{1}{\omega C}$$

giving
$$1 + \frac{Z_1}{2Z_2} = 1 + \frac{j\omega L}{-2(1/\omega C)} = 1 - \frac{\omega^2 LC}{2}$$

or
$$\cosh \alpha . \cos \beta + j \sinh \alpha . \sin \beta = 1 - \frac{\omega^2 LC}{2}$$

Hence we have
$$\cosh \alpha . \cos \beta = 1 - \frac{\omega^2 LC}{2}$$
$$\sinh \alpha . \sin \beta = 0$$

Pass band

Here $\alpha = 0$ with $\cosh \alpha = 1$ and $\sinh \alpha = 0$ with

$$\cos \beta = 1 - \frac{\omega^2 LC}{2}$$

Since $-1 < \cos \beta < +1$, the two limiting conditions are given by

(a)
$$-1 = 1 - \frac{\omega^2 LC}{2}$$

and
$$\omega^2 = \frac{4}{LC}$$

or $f = 1/\pi\sqrt{LC}$ which is the upper limit or cut-off frequency f_c obtained previously

(b)
$$+1 = 1 - \frac{\omega^2 LC}{2}$$

and
$$\omega^2 = 0$$

or
$$f = 0 \text{ which is the lower limit}$$

The pass band therefore extends from 0 to f_c as shown previously.

Phase shift

Since
$$\cos \beta = 1 - \frac{\omega^2 LC}{2} = 1 - 2\left(\frac{f}{f_c}\right)^2$$

with
$$\beta = \cos^{-1}\left[1 - 2\left(\frac{f}{f_c}\right)^2\right]$$

and is plotted in Fig. 36.

Stop-band

Since β varies from 0 to π over the pass band, at f_c and beyond, it has the maximum value π.

i.e.
$$\cos \beta = \cos \pi = -1$$

and
$$\cosh \alpha = \frac{\omega^2 LC}{2} - 1 = 2\left(\frac{f}{f_c}\right)^2 - 1$$

or
$$\alpha = \cosh^{-1}\left[2\left(\frac{f}{f_c}\right)^2 - 1\right]$$

and is plotted in Fig. 36.

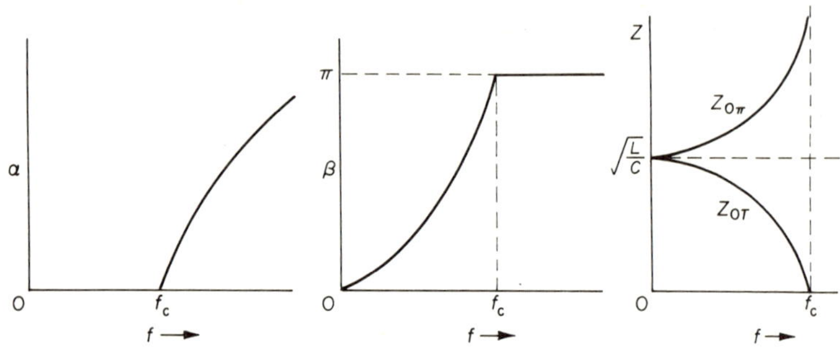

Fig. 36

Impedance considerations

For a T network Z_{0T} is given by

$$Z_{0T} = \sqrt{Z_1 Z_2 + \frac{Z_1^2}{4}}$$

with
$$Z_1 = j\omega L \quad \text{and} \quad Z_1 Z_2 = \frac{L}{C}$$

$$Z_2 = -j\frac{1}{\omega C}$$

Hence
$$Z_{0T} = \sqrt{\frac{L}{C} - \frac{\omega^2 L^2}{4}}$$

at any angular frequency ω.

When
$$\omega = 0$$

$$Z_{0T} = \sqrt{\frac{L}{C}}$$

When
$$\omega = \omega_c = \frac{2}{\sqrt{LC}}$$

$$Z_{0T} = \sqrt{\frac{L}{C} - \frac{L}{C}} = 0$$

Similarly
$$Z_{0\pi} = \frac{Z_1 Z_2}{Z_{0T}} = \frac{L/C}{\sqrt{(L/C) - (\omega^2 L/4)}}$$

When
$$\omega = 0$$

$$Z_{0\pi} = \sqrt{\frac{L}{C}}$$

When
$$\omega = \omega_c$$

$$Z_{0\pi} = \frac{L/C}{\sqrt{(L/C) - (L/C)}} = \frac{L/C}{0} = \infty$$

These results are plotted in Fig. 36 and summarize the important characteristics of the filter.

Design notes

1. The basic equations used are the design resistance $R_0 = \sqrt{L/C}$ and cut-off frequency $f_c = 1/\pi\sqrt{LC}$. These equations yield

$$L = CR_0^2, \qquad C = \frac{1}{\pi R_0 f_c}$$

2. The component values for either a T or π network are then obtained using the standard configuration given in Fig. 34.

EXAMPLE 10

Each of the series arms of a symmetrical T-type low-pass filter section consists of an inductor of 18 mH having negligible resistance while the shunt arm is a $0.1\ \mu$F capacitor. Indicate how the characteristic impedance varies with frequency and calculate its value at 1 kHz and 8 kHz Derive any formulae used.

What are the limitations of such a filter section? Indicate how the performance can be improved. L.U.B.Sc(Eng) Tels., 1963

Solution

Here
$$L = 18\ \text{mH} + 18\ \text{mH} = 36\ \text{mH}$$

$$C = 0.1\ \mu\text{F}$$

The characteristic impedance of a T-type network is given by

$$Z_{OT} = \sqrt{Z_1 Z_2 + \frac{Z_1^2}{4}}$$

Here
$$Z_1 = j\omega L, \qquad Z_2 = -j\frac{1}{\omega C}$$

and
$$Z_{OT} = \sqrt{\frac{L}{C} - \frac{\omega^2 L^2}{4}}$$

At
$$f = 0$$

$$Z_{OT} = \sqrt{\frac{L}{C}}$$

At other frequencies, Z_{OT} decreases to zero when

$$\frac{L}{C} = \frac{\omega^2 L^2}{4}$$

or
$$f = \frac{1}{\pi\sqrt{LC}}$$

which is the cut-off frequency of the filter. The variation of Z_{OT} is plotted in Fig. 36.

At 1 kHz

$$Z_{OT} = \sqrt{\frac{36 \cdot 10^{-3}}{0{\cdot}1 \times 10^{-6}} - \frac{(2\pi \cdot 10^3)^2 \cdot 36^2 \cdot 10^{-6}}{4}}$$

$$= \sqrt{36 \cdot 10^4 - \pi^2 \cdot (36)^2}$$

$$= 60\sqrt{96 \cdot 45}$$

$$= 588\ \Omega$$

At 8 kHZ

$$Z_{OT} = \sqrt{\frac{36 \cdot 10^{-3}}{0{\cdot}1 \times 10^{-6}} - \frac{(2\pi \times 8 \times 10^3)^2 \cdot (36)^2 \cdot 10^{-6}}{4}}$$

$$= \sqrt{36 \cdot 10^4 - \pi^2 \cdot 64 \cdot (36)^2}$$

$$= j60\sqrt{9{\cdot}86 \times 6{\cdot}4 \times 3{\cdot}6 - 100}$$

$$= j60\sqrt{127}$$

$$= j676\ \Omega$$

Two serious limitations of the filter are:

1. The attenuation rises slowly beyond cut-off.
2. The impedance varies considerably over the pass-band, causing mismatch.

These drawbacks are reduced by using an *m*-derived filter having the same characteristic impedance as the prototype. By using an *m*-derived filter whose *m* value is about 0·3 in cascade with the prototype, the attenuation curve can be made to rise sharply after cut-off. Furthermore, since the input impedance of a half-section derived filter with $m = 0·6$ is fairly constant over the pass-band, two such matching half-sections can be used at the input and output ends, respectively, of the cascaded prototype and *m*-derived sections as illustrated in Fig. 40.

EXAMPLE 11
Determine, from first principles, the element values required for a constant-*k*, T-section, low-pass filter, suitable for insertion in a 600 Ω circuit. The cut-off frequency is to be 1 MHz. Determine the overall attenuation in db, at a frequency of 2 MHz, of two such sections connected in cascade and terminated with an ideal load equal to the characteristic impedance.

Solution
Let the filter have the configuration shown in Fig. 34.

Since
$$Z_{0T} = \sqrt{\frac{L}{C}} = 600$$

Hence
$$\frac{L}{C} = 36 \cdot 10^4 \tag{1}$$

Also
$$f_c = \frac{1}{\pi\sqrt{LC}} = 10^6$$

or
$$LC = \frac{1}{\pi^2 \cdot 10^{12}} \tag{2}$$

From equations (1) and (2) we obtain
$$L^2 = \frac{36 \cdot 10^4}{\pi^2 \cdot 10^{12}}$$

or
$$L = \frac{6}{\pi} \cdot 10^{-4} = 0.191 \text{ mH}$$

and
$$C = \frac{1}{\pi^2 \cdot 10^{12} \cdot (0.191) \cdot 10^{-3}} = 530 \text{ pf}$$

Now
$$\cosh \gamma = 1 + \frac{Z_1}{2Z_2} \quad \text{with} \quad Z_1 = j\omega L$$

$$Z_2 = -j\frac{1}{\omega C}$$

$$= 1 + \frac{j\omega L}{2[-j(1/\omega C)]}$$

$$= 1 - \frac{\omega^2 LC}{2}$$

Hence
$$\cosh \alpha \cdot \cos \beta = 1 - \frac{\omega^2 LC}{2}$$

Now the frequency of 2 MHz is in the stop-band since the cut-off frequency is 1 MHz, and so $\beta = \pi$

Hence
$$\cosh \alpha \cdot \cos \pi = 1 - \frac{\omega^2 LC}{2}$$

and
$$\cosh \alpha = \frac{\omega^2 LC}{2} - 1$$

$$= 2\left(\frac{f}{f_c}\right)^2 - 1$$

$$= 2\left[\frac{2 \times 10^6}{1 \times 10^6}\right] - 1$$

$$= (2 \times 4) - 1 = 7$$

and
$$\alpha = \cosh^{-1} 7 = 2.64 \text{ nepers} = 22.8 \text{ db}$$

or
$$\alpha = 45.6 \text{ db for two sections in cascade}$$

EXAMPLE 12

What is meant by the statement that the phase-shift coefficient at a frequency f across a low-pass constant-k filter with a cut-off frequency f_0 is given by $2 \sin^{-1}(f/f_0)$ radians per section?

Derive an expression for the time delay that such a filter would impose on an applied attenuating voltage of frequency f assuming the filter to be terminated in its characteristic impedance.

An artificial line is constructed from twenty similar sections with a nominal characteristic impedance of 2000 Ω to provide at very low frequencies ($f \ll f_0$) a total phase delay of 2 μs. Calculate the cut-off frequency and the component values of each section.

(Free space velocity of light $= 3 \times 10^8$ m/s.)

L.U.B.Sc(Eng) Tels., 1965

Solution

When a signal of frequency f passes through a low-pass filter, it suffers a phase-shift of β rads/section, i.e. the output lags the input by this amount.

This phase lag is a function of frequency and usually increases with f. The exact relation is given by the expression quoted viz

$$\beta = 2 \sin^{-1}\left(\frac{f}{f_0}\right)$$

or
$$\sin\frac{\beta}{2} = \frac{f}{f_0}$$

where f is the frequency considered and f_0 is the cut-off frequency of the filter.

If β is the phase delay and t_d is the time delay, then by simple proportion

$$\frac{\beta}{2\pi} = \frac{t_d}{T} \text{ where } T \text{ is the periodic time}$$

or
$$t_d = \frac{\beta}{2\pi} \cdot T = \frac{\beta}{2\pi f}$$

or
$$t_d = \frac{\beta}{\omega} = \frac{1}{\pi f}\sin^{-1}\left(\frac{f}{f_0}\right)$$

PROBLEM

Time delay/section $= 2 \times 10^{-6}/20 = 10^{-7}$ s.

Hence
$$\frac{\beta}{\omega} = 10^{-7}$$

or
$$\frac{\omega}{\beta} = 10^7 = v_\text{p},$$

the phase velocity along the line.

Now
$$v_\mathrm{p} = \frac{1}{\sqrt{LC}} \text{ for an artificial line}$$

and
$$R_0 = \sqrt{\frac{L}{C}} = 2000$$

Hence
$$\frac{1}{\sqrt{LC}} = 10^7$$

$$\sqrt{\frac{L}{C}} = 2000$$

This leads to
$$C = 0.5 \times 10^{-10} = 50 \text{ pf}$$

and
$$L = 4 \times 10^6 \times C = 0.2 \text{ mH}$$

Also
$$f_\mathrm{c} = \frac{1}{\pi\sqrt{LC}} = \frac{1}{\pi\sqrt{0.2 \times 10^{-3} \times 50 \times 10^{-12}}}$$

or
$$f_\mathrm{c} = 3.18 \text{ MHz}$$

(b)   High-pass filter

It passes all frequencies *above* a certain cut-off value f_c. The configuration and symbolism is shown in Fig. 37.

For the T-section we have

$$Z_1 = -j\frac{1}{\omega C} \quad \text{or} \quad X_1 = -\frac{1}{\omega C}$$

$$Z_2 = j\omega L \qquad\qquad X_2 = \omega L$$

and
$$\frac{X_1}{4} + X_2 = \omega L - \frac{1}{4\omega C}$$

A plot of X_1 and $[(X_1/4) + X_2]$ is shown in Fig. 38 and the cut-off frequency occurs when $Z_{0T} = 0$, i.e. it changes between real and reactive values.

At cut-off when $Z_{0T} = 0$ we have

$$\omega L - \frac{1}{4\omega C} = 0$$

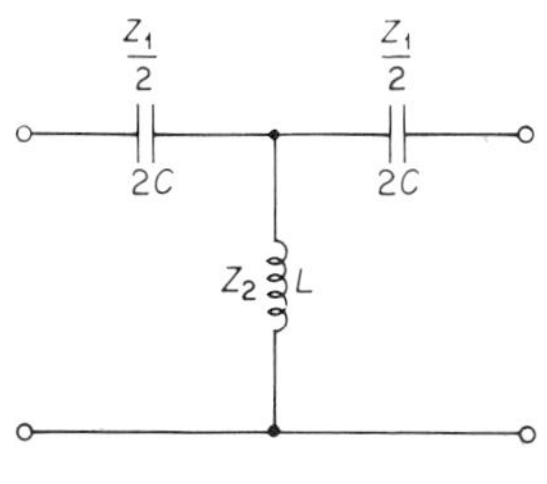 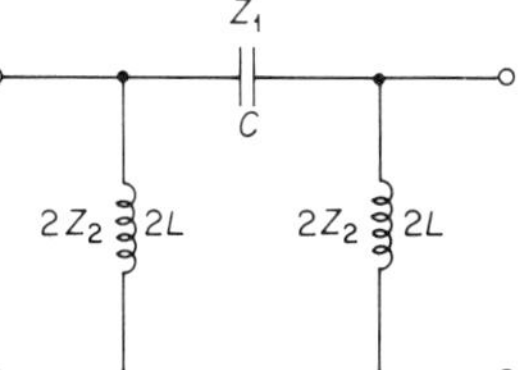

Fig. 37

or

$$\omega^2 = \frac{1}{4LC}$$

and

$$f_c = \frac{1}{4\pi\sqrt{LC}}$$

which is true for both the T or π form.

Now previously we obtained

$$\cosh \gamma = 1 + \frac{Z_1}{2Z_2} \text{ for a T or } \pi \text{ network}$$

Here

$$Z_1 = -j\frac{1}{\omega C}$$

$$Z_2 = j\omega L$$

So

$$\cosh \gamma = 1 + \frac{[-j(1/\omega C)]}{2j\omega L} = 1 - \frac{1}{2\omega^2 LC}$$

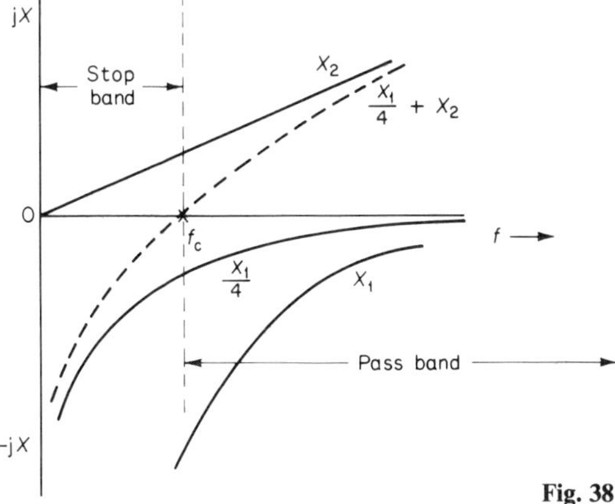

Fig. 38

Hence
$$\cosh \alpha \cdot \cos \beta = 1 - \frac{1}{2\omega^2 LC}$$

$$\sinh \alpha \cdot \sin \beta = 0$$

Pass band $(\alpha = 0)$
With $\alpha = 0$ $\cosh \alpha = 1$ and we obtain

$$\cos \beta = 1 - \frac{1}{2\omega^2 LC}$$

Using extreme limits ± 1 for $\cos \beta$ gives

(a)
$$-1 = 1 - \frac{1}{2\omega^2 LC}$$

or
$$\omega^2 = \frac{1}{4LC}$$

and $f_1 = 1/4\pi\sqrt{LC}$ which is the value f_c obtained above.

(b)
$$+1 = 1 - \frac{1}{2\omega^2 LC}$$

or
$$\omega^2 = \frac{1}{0} = \infty$$

or $f_2 = \infty$ which is the upper limit.
Hence, the pass band extends from f_1 to ∞.

Phase shift
We have
$$\cos \beta = 1 - \frac{1}{2\omega^2 LC} = 1 - 2\left(\frac{f_c}{f}\right)^2 \quad \text{since} \quad f_c = \frac{1}{4\pi\sqrt{LC}}$$

Hence
$$\beta = \cos^{-1}\left[1 - 2\left(\frac{f_c}{f}\right)^2\right]$$

and is plotted in Fig. 39.

Stop band
Since β varies from 0 to $-\pi$ over the pass-band, at f_c and below it, β has
the maximum value $-\pi$,

giving
$$\cos \beta = \cos(-\pi) = -1$$

Hence
$$\cosh \alpha . \cos \beta = \cosh \alpha . (-1) = 1 - \frac{1}{2\omega^2 LC}$$

or
$$\cosh \alpha = \frac{1}{2\omega^2 LC} - 1 = 2\left(\frac{f_c}{f}\right)^2 - 1$$

with
$$\alpha = \cosh^{-1}\left[2\left(\frac{f_c}{f}\right)^2 - 1\right]$$

and is plotted in Fig. 39

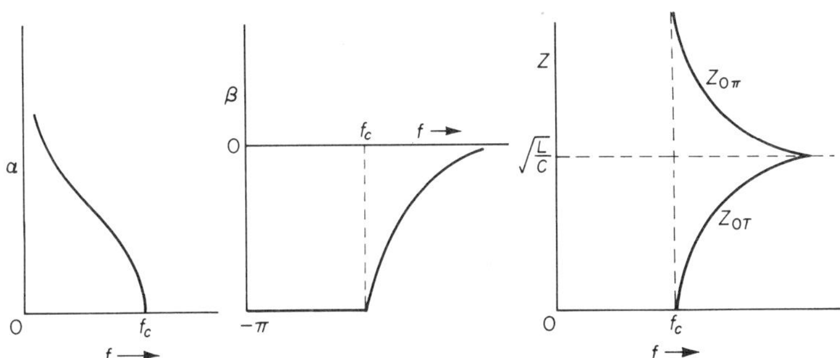

Fig. 39

Impedance considerations

$$Z_{0T} = \sqrt{Z_1 Z_2 + \frac{Z_1^2}{4}} \quad \text{where} \quad Z_1 = -j\frac{1}{\omega C}$$

$$Z_2 = j\omega L$$

or
$$Z_{0T} = \sqrt{\frac{L}{C} - \frac{1}{4\omega^2 C^2}}$$

At $\omega = \infty$

$$Z_{0T} = \sqrt{\frac{L}{C}}$$

At $\omega = \omega_c$

$$Z_{0T} = 0$$

Once again Z_{0T} varies considerably over the pass band from f_c to ∞.

Similarly,
$$Z_{0\pi} = \frac{Z_1 Z_2}{Z_{0T}} = \frac{L/C}{\sqrt{(L/C) - (1/4\omega^2 C^2)}}$$

At $\omega = \infty$,

$$Z_{0\pi} = \sqrt{\frac{L}{C}}$$

At $\omega = \omega_c$,

$$Z_{0\pi} = \frac{L/C}{0} = \infty$$

With the above results we obtain the three important characteristics of the filter as in Fig. 39.

EXAMPLE 13

Design a prototype (constant-k), T-section, high-pass filter to have cut-off frequency of $10,000/2\pi$ Hz and a characteristic impedance of 500 Ω at a very high frequency.

The equations

$$\cosh \gamma = 1 + \frac{Z_1}{2Z_2}$$

and

$$Z_k = \left[Z_1 Z_2 + \frac{Z_1^2}{4} \right]^{1/2}$$

may be assumed, where γ = the propagation constant, Z_1 = total series impedances, Z_2 = shunt impedance and Z_k = characteristic impedance of a T section.

Explain your solution.

Discuss the imperfections of the prototype and describe briefly how they may be overcome.   L.U.B.Sc(Eng) Tels., 1960

Solution

Since the equations for $\cosh \gamma$ and Z_k are given the solution must be obtained from these results.

The configuration for a T section is shown in Fig. 37

where

$$Z_k = \sqrt{Z_1 Z_2 + \frac{Z_1^2}{4}} = \sqrt{\frac{L}{C} - \frac{1}{4\omega^2 C^2}}$$

The design impedance used is the value

$$Z_k = \sqrt{\frac{L}{C}} \tag{1}$$

Now $\qquad \cosh \gamma = 1 + \dfrac{Z_1}{2Z_2} = 1 - \dfrac{j(1/\omega C)}{2j\omega L} = 1 - \dfrac{1}{2\omega^2 LC}$

Hence $\qquad\qquad \cosh \alpha . \cos \beta = 1 - \dfrac{1}{2\omega^2 LC}$

In the pass band $\alpha = 0$ with $\cosh \alpha = 1$ and $-1 < \cos \beta < +1$

Hence $\qquad\qquad\qquad -1 = 1 - \dfrac{1}{2\omega^2 LC}$

or $\qquad\qquad\qquad\qquad \omega^2 - \dfrac{1}{4LC}$

or $\qquad\qquad\qquad\qquad f_c = \dfrac{1}{4\pi\sqrt{LC}}$ $\qquad\qquad$ (2)

From (1) and (2)

$$\frac{Z_k}{f_c} = 4\pi . L$$

or $\qquad L = \dfrac{Z_k}{4\pi f_c} = \dfrac{500}{4\pi . (10,000/2\pi)} = 25 \text{ mH}$

and $\qquad C = \dfrac{L}{Z_k^2} = \dfrac{1}{40 . (500)^2} = 0.1 \text{ } \mu\text{F}$

The filter structure is shown in Fig. 40.
The imperfections of the prototype above are:

1. The value of Z_k varies over the pass band.
2. The attenuation α rises slowly after f_c.

The first imperfection is reduced by using a matching m-derived half section with $m = 0.6$ at the input to the prototype, since the input impedance remains fairly constant over most of the pass band. The second imperfection is reduced by using an m-derived full-section with $m = 0.3$ (approx.) after the prototype section, since its attenuation rises steeply at f_c. For final matching, this section is terminated by another m-derived half-section with $m = 0.6$.

This is illustrated in Fig. 40

EXAMPLE 14
Determine, from first principles, the element values required for a constant-k, T-section high-pass filter suitable for insertion in a 75 Ω

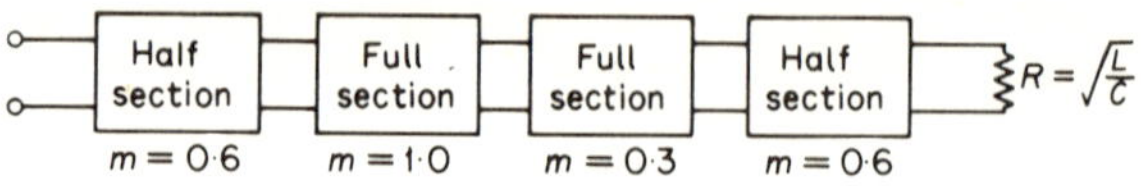

Fig. 40

circuit. The cut-off frequency is to be 4 MHz. Sketch the amplitude and phase response of such a section when correctly terminated. Determine the frequency range over which the characteristic impedance is resistive and within 10% of 75 Ω.   L.U.B.Sc(Eng) Tels., 1967

Solution

Here
$$Z_1 = -j\,\frac{1}{\omega C}$$

$$Z_2 = j\omega L$$

The design equations are

$$Z_0 = \sqrt{\frac{L}{C}} = 75\ \Omega$$

$$f_c = \frac{1}{4\pi\sqrt{LC}} = 4 \times 10^6$$

or
$$\frac{L}{C} = 75 \times 75$$

and
$$\frac{1}{LC} = 256\pi^2 \times 10^{12}$$

Hence
$$\frac{L/C}{1/LC} = L^2 = \frac{75 \times 75}{256\pi^2 \times 10^{12}}$$

or
$$L = 1{\cdot}495\ \mu\text{H}$$

and
$$C = \frac{L}{75 \times 75} = \frac{1{\cdot}495 \times 10^{-6}}{75 \times 75} = 266\ \text{pf}$$

Over the pass band, $\alpha = 0$

Hence
$$\left|\frac{V_1}{V_2}\right| = e^0 = 1 \quad \text{or} \quad V_2 = V_1$$

Within the attenuation band, V_2 falls eventually to 0.

NOTE. The amplitude response was asked for, and this is strictly $|V_2/V_1|$. Alternatively, the attenuation curve may have been required and is shown in Fig. 39.

Again
$$Z_{0T} = R_0\sqrt{1 - \left(\frac{f_c}{f}\right)^2}$$

or
$$(0{\cdot}9)^2 = 1 - \left(\frac{4}{f}\right)^2 \quad \text{where } f \text{ is in MHz}$$

and
$$f = 9{\cdot}2 \text{ MHz}$$

The frequency range asked for extends from 9·2 MHz to ∞.

(c)　Constant-k filters

The low-pass and high-pass filters considered so far are characterized by the fact that

$$Z_1 Z_2 = \frac{L}{C} \quad \text{(a constant)}$$

for the low-pass or high-pass filter.

This constant is usually denoted by k^2, hence the name. In fact, k is the design impedance of the filter.

5
m-derived filters

Two disadvantages of simple constant-k filters are that (a) the attenuation characteristic does not rise sharply after cut-off and (b) the characteristic impedance Z_0 is not constant in the pass band but varies considerably.

These drawbacks can more or less be minimized by using *m*-derived filters where m is a numerical factor (between 0 and 1) and the value of m determines the property of the *m*-derived filter. Essentially, the component values of the prototype filter are modified by the factor m, and an additional component has to be used in order to maintain the same value of Z_0 regardless of the value of m.

5.1   Low-pass T filter

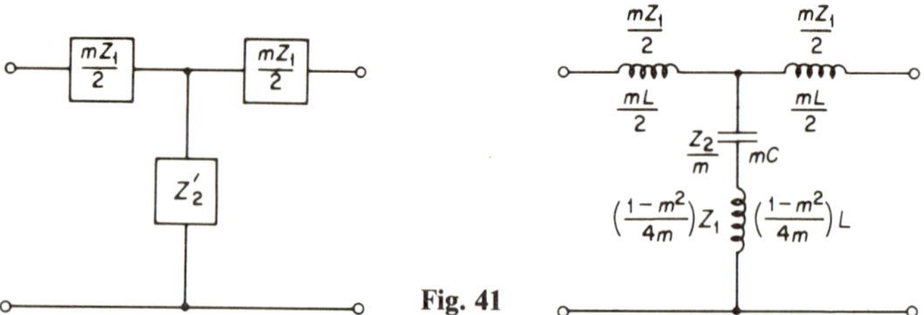

Fig. 41

Consider the *m*-derived filter in Fig. 41 as having total series and shunt impedances mZ_1 and Z'_2 respectively, such that both the prototype and *m*-derived filters have the same characteristic impedance Z_{0T}.

Hence
$$Z_{0T} = \sqrt{Z_1 Z_2 + \frac{Z_1^2}{4}} = \sqrt{mZ_1 Z'_2 + m^2 \frac{Z_1^2}{4}}$$

or
$$Z_1 Z_2 + \frac{Z_1^2}{4} = mZ_1 Z'_2 + m^2 \frac{Z_1^2}{4}$$

giving
$$Z'_2 = \frac{Z_2}{m} + \left(\frac{1 - m^2}{4m}\right) Z_1$$

This result means that Z_{0T} will be the same for the *m*-derived filter provided the shunt impedance Z'_2 consists of an inductance $[(1 - m^2)/4m]Z_1$

in series with a capacitance Z_2/m. Hence an additional component has to be used in the shunt arm.

5.2 Low-pass π filter

To maintain the same value of $Z_{0\pi}$ for both filters the series arm must be Z'_1 and the shunt arms $2Z_2/m$ such that

$$Z_{0\pi} = \frac{Z_1 Z_2}{\sqrt{Z_1 Z_2 + (Z_1^2/4)}} = \frac{Z'_1 Z_2/m}{\sqrt{(Z'_1 Z_2/m) + [(Z'_1)^2/4]}}$$

or

$$\frac{Z_1}{Z_1 Z_2 + (Z_1^2/4)} = \frac{(Z'_1)^2/m^2}{(Z'_1 Z_2/m) + [(Z'_1)^2/4]}$$

or

$$4Z_1 Z_2 m + m^2 Z_1 Z'_1 = 4Z_2 Z'_1 + Z_1 Z'_1$$

and

$$Z'_1 = \frac{4mZ_1 Z_2}{4Z_2 + (1 - m^2)Z_1}$$

which amounts to an inductive impedance mZ_1 in parallel with a capacitive impedance of $4mZ_2/(1 - m^2)$.

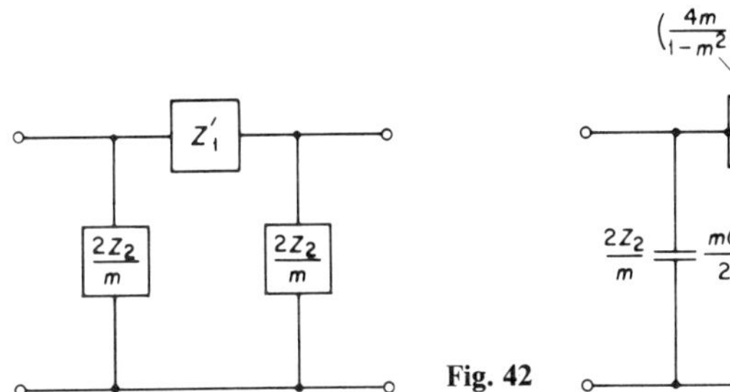

Fig. 42

5.3 Characteristics of the *m*-derived filter

In the case of the T filter the shunt arm resonates at a certain frequency and short-circuits the input signal, i.e. it has infinite attenuation. This frequency is called the frequency of infinite attenuation f_∞.

Hence

$$f_\infty = \frac{1}{2\pi\sqrt{[(1 - m^2)/4m]L \cdot mC}} = \frac{1}{\pi\sqrt{LC}\sqrt{1 - m^2}}$$

or

$$f_\infty^2 = \frac{f_c^2}{1 - m^2} \quad \text{where} \quad f_c = \frac{1}{\pi\sqrt{LC}}$$

and

$$f_\infty = \frac{f_c}{\sqrt{1 - m^2}}$$

or

$$m = \sqrt{1 - \left(\frac{f_c}{f_\infty}\right)^2}$$

This shows that the cut-off frequency can be made to approach close to the frequency of infinite attenuation by making m small; in practice it is usually about 0·3. The values of α and β for $m = 0·3$ are shown in Fig. 43.

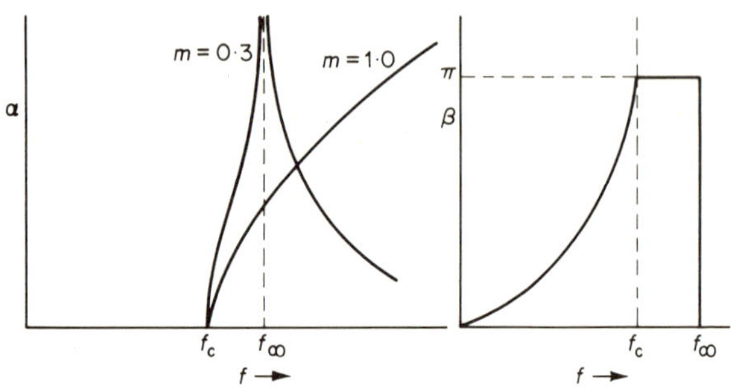

Fig. 43

5.4   High-pass filter

Similar considerations apply to this case and the corresponding configurations are shown in Fig. 44. The value of m is given by

$$m = \sqrt{1 - \left(\frac{f_\infty}{f_c}\right)^2}$$

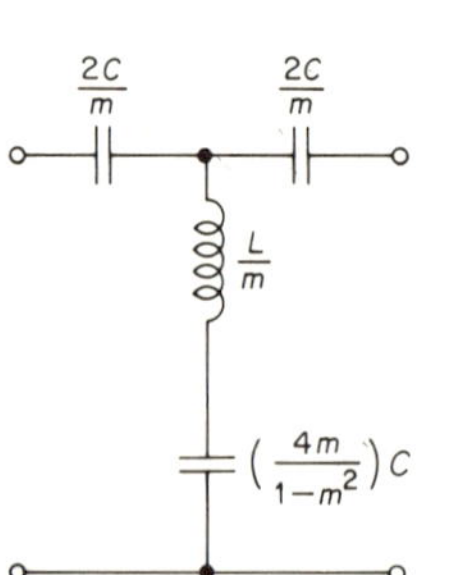

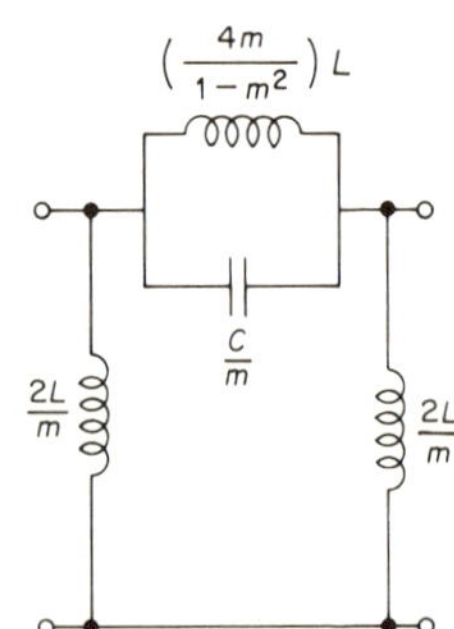

Fig. 44

5.5 Terminating half-sections

The second draw-back of the prototype filter can be minimized by using half an *m*-derived T or π section. Such half-sections have the property of having a fairly constant value of Z_0 over most of the pass-band if $m = 0.6$. It is usual to use two such sections (one at the input and one at the output) and are known as terminating half-sections.

The input and output impedances of the half-section are Z_{0T} and $Z_{0\pi m}$ respectively. Z_{0T} is by definition independent of the value of *m* but $Z_{0\pi m}$ is a function of *m* and the variation is shown in Fig. 45. These impedances are image impedances.

In this case $Z_{0\pi}$ is independent of *m* by definition, but Z_{0Tm} varies with *m* and is shown in Fig. 46.

Since such half-sections present image impedances at either end, of value Z_{0T} or $Z_{0\pi}$, they can be used for matching a T network to a π network or vice versa.

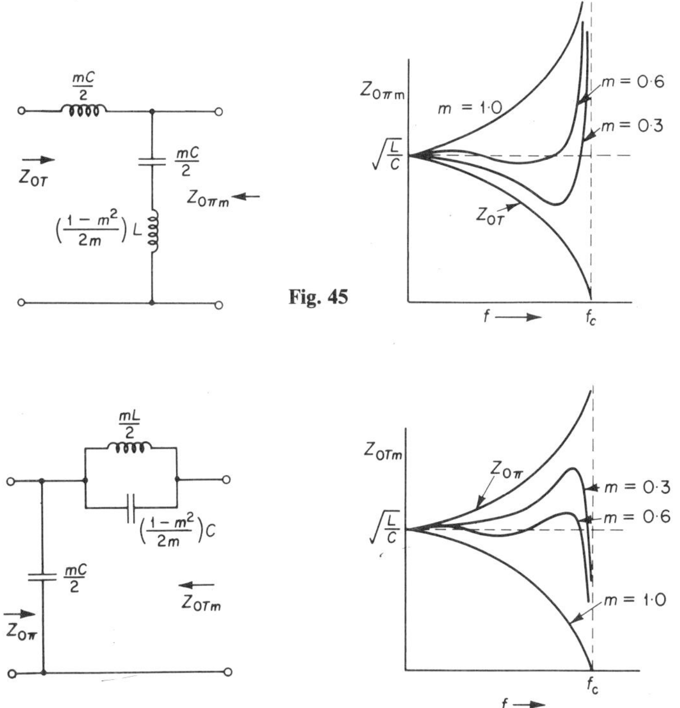

Fig. 45

Fig. 46

Comment

Similar considerations apply to the corresponding high-pass terminating half-sections.

5.6 Composite filters

Several *m*-derived filters with different values of *m* can be connected in tandem to form a composite filter, since they all have the same characteristic impedance, which is independent of *m*.

A typical structure would consist of the prototype ($m = 1 \cdot 0$) in series with one or two sections having a value of $m = 0 \cdot 3$ or $m = 0 \cdot 4$. The *m*-derived sections improve the attenuation *immediately after* cut-off while the prototype improves it *well after* cut-off. Furthermore, terminating half-sections with $m = 0 \cdot 6$ are used at either end to obtain constant match over the pass band.

EXAMPLE 15

Design a composite low-pass filter with $f_c = 1000$ Hz, $f_\infty = 1200$ Hz and $Z_0 = 600\ \Omega$. It must consist of a prototype T section, one *m*-derived T section and two terminating half-sections.

Solution

The prototype is designed first with $m = 1 \cdot 0$, then the *m*-derived half-section with the appropriate value of *m* and finally the terminating half-sections with $m = 0 \cdot 6$.

Prototype.   The design equations are

$$Z_0 = \sqrt{\frac{L}{C}} = R_0$$

$$f_c = \frac{1}{\pi\sqrt{LC}}$$

Hence
$$L^2 = \frac{R_0^2}{\pi^2 f_c^2} \quad \text{or} \quad L = \frac{R_0}{\pi f_c}$$

and
$$C = \frac{1}{\pi f_c R_0}$$

This yields
$$L = \frac{600}{\pi \times 10^3} = 0 \cdot 191 \text{ H}$$

$$C = \frac{1}{\pi \cdot 10^3 \cdot 600} = 0 \cdot 53\ \mu\text{F}$$

m-derived full-section.   Here

$$m = \sqrt{1 - \left(\frac{f_c}{f_\infty}\right)^2} = \sqrt{1 - \frac{1000}{1200}} = 0{\cdot}553$$

and
$$mL = 0{\cdot}553 \times 0{\cdot}191 = 0{\cdot}1055 \text{ H}$$

$$\left(\frac{1 - m^2}{4m}\right)L = \left[\frac{1 - (0{\cdot}553)^2}{4 \times 0{\cdot}553}\right] \times 0{\cdot}191 = 0{\cdot}06 \text{ H}$$

$$mC = 0{\cdot}553 \times 0{\cdot}53 \ \mu\text{F} = 0{\cdot}293 \ \mu\text{F}$$

Half-section.   Here
$$m = 0{\cdot}6$$

$$\frac{mZ_1}{2} = \frac{mL}{2} = \frac{0{\cdot}6}{2} \times 0{\cdot}191 = 0{\cdot}0573 \text{ H}$$

$$\frac{2Z_2}{m} = \frac{mC}{2} = \frac{0{\cdot}6}{2} \times 0{\cdot}53 = 0{\cdot}159 \ \mu\text{F}$$

$$\left(\frac{1 - m^2}{2m}\right)Z_1 = \left(\frac{1 - m^2}{2m}\right)L = \left[\frac{1 - (0{\cdot}6)^2}{2 \times 0{\cdot}6}\right] \times 0{\cdot}191 = 0{\cdot}1018 \text{ H}$$

The component values of the basic configuration are shown in Fig. 47(a) and those of the final configuration in Fig. 47(b).

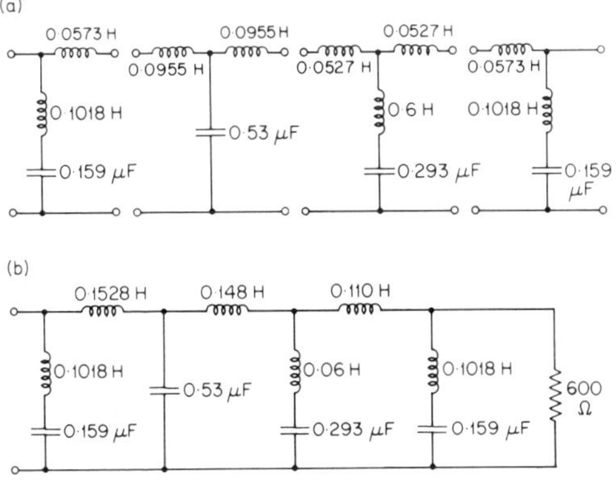

Fig. 47

6
Modern filter theory

6.1 Introduction

In the past, filters have been designed on the image-parameter basis as described in the previous two chapters. More recently however, due to the advent of the computer, which can undertake tedious calculations, filter design is based on the insertion-loss or synthesis approach. The essential difference is that in the older theory, a filter was designed according to the theory and its characteristics were accepted as final. In the modern approach, the filter is built up or synthesized bit by bit to give the required characteristics as nearly as possible.

This modern method is far more powerful and flexible than the older classical theory. The modern design procedure is to synthesize the filter from its voltage transfer function as nearly as possible, through the use of *approximation theory*.[7] In order to build a practical filter, only certain types of functions are physically realizable and are known as rational functions. The type of filter designed depends on the rational function chosen.

The three most common types of filters are the Butterworth, Chebyshev and elliptic function filters. The procedure is to choose one of the realizable rational functions and to obtain the required transfer function from the given amplitude or phase characteristic. The transfer function is then divided into various parts and the filter synthesized according to each part.

6.2 Design procedure

If $V_1(s)$ and $V_2(s)$ are the voltages at the input and output of the filter, then the voltage transfer function $F(s)$ which is considered to have only poles is given by

$$F(s) = \frac{V_2(s)}{V_1(s)} = \frac{1}{V_1(s)/V_2(s)} = \frac{1}{(s + s_1)(s + s_2)}$$

where $s = j\omega$ and s_1, s_2 etc. are the required poles.

In an ideal low-pass filter, for example, the ideal response would be unity over the pass band and would fall to zero at the cut-off frequency

f_c. Such a characteristic could never be obtained in practice but may be approximated by various functions whose behaviour is best examined in terms of its poles.

The basic procedure is to choose the appropriate $F(s)$ and from it obtain the conjugate $F(-s)$. Multiplication of these functions yields $|F(s)|^2$ which involves the square of the amplitude only and its poles are then determined. Since the physical function is $F(s)$, only poles in the left half plane are considered in the design. This then enables $F(s)$ to be synthesized as a network.

Such filters do not possess a definite cut-off frequency as in classical theory, but are usually 'normalized' at the convenient value of $\omega = 1$, which is specified by the designer. Butterworth filters are usually normalized at the 3-db point, and in the Chebyshev filter, this is taken as the last point of interest in the pass band. The analysis is best illustrated using the following designs.

6.3 Butterworth filter[8-10]

The Butterworth approximation of the ideal low-pass filter response is shown in Fig. 48 and given by

$$F_n(s) \cdot F_n(-s) = |F_n(s)|^2 = \frac{1}{1 + \omega^{2n}} = \frac{1}{1 + (s/j)^{2n}}$$

where $n = 1, 2, 3$ etc. and gives the degree of the filter.

The poles are given by $(s/j)^{2n} + 1 = 0$

i.e.
$$\left(\frac{s}{j}\right)^{2n} = -1$$

or
$$(-1)^n \cdot s^{2n} = -1 = e^{j(2k-1)\pi}$$

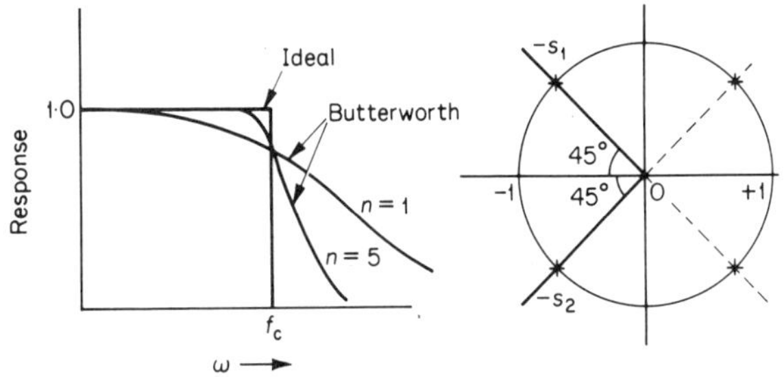

Fig. 48

where $k = 1, 2$ etc. . . . .

Hence
$$s^{2n} = e^{j[2k+n-1]\pi}$$

or
$$s_k = e^{j[(2k+n-1)/2n]\pi}$$

and the various poles are s_1, s_2 etc. . . . .

Hence
$$F(s) = \frac{1}{(s + s_1)(s + s_2)\cdots}$$

As an example, assume $n = 2$ and we have

$$s^4 = -1 = 1\underline{/180}$$

or
$$s^2 = \pm 1\underline{/90} = 1\underline{/\pm 90}$$

and
$$s = \pm 1\underline{/\pm 45}$$

i.e. the required poles lie on a unit circle in the left-half plane at $\underline{/s} \pm 45°$ as shown in Fig. 48.

$$F(s) = \frac{1}{(s + 1\underline{/45})(s + 1\underline{/-45})}$$

$$= \frac{1}{(s^2 + 1\cdot414s + 1)}$$

Values of $F(s)$ for different n are shown in Table 2.

Table 2

n	$F(s)$
1	$\dfrac{1}{s + 1}$
2	$\dfrac{1}{s^2 + 1\cdot414s + 1}$
3	$\dfrac{1}{s^3 + 2s^2 + 2s + 1}$
4	$\dfrac{1}{s^4 + 2\cdot613s^3 + 3\cdot414s^2 + 2\cdot613s + 1}$

A plot of $F(s)$ for various values of n is shown in Fig. 48. The amplitude response, because of its initial flatness, is referred to as the 'maximally flat' response over the pass band and is monotonic. In the attenuation band it falls off as ω^{-n} and the roll-off in db is given by

$$20 \log_{10} F(s) = 20 \log_{10} \frac{1}{\sqrt{1 + \omega^{2n}}}$$

$$= 20 \log_{10} \omega^{-n}$$

$$= -20n \log_{10} \omega$$

i.e. the maximum value of db 'roll-off' is $6n$db/octave or $20n$/decade.

Filters are usually designed using 'normalized' values as mentioned earlier, i.e. for $\omega_c = 1$, $R_0 = 1\,\Omega$. The answers obtained are then suitably scaled up to give the exact physical values of the practical filter. For example, if R_0, L, C are the unnormalized filter parameters and R_n, L_n, C_n are the normalized values, we have that the actual frequency s is related to the normalized frequency s_n by the relation

$$s_n = \frac{s}{\omega_c}$$

where ω_c is taken as the normalizing cut-off frequency.

If only the frequency is normalized but not R_0, then the other impedances must be the same at the appropriate frequency and so we have

$$sL = s_n L_n$$

or

$$L = \frac{s_n L_n}{s} = \frac{L_n}{\omega_c}$$

and

$$\frac{1}{sC} = \frac{1}{s_n C_n}$$

or

$$C = \frac{s_n C_n}{s} = \frac{C_n}{\omega_c}$$

If, furthermore, the level of resistance R of the filter is normalized also, so that

$$R_n = \frac{R}{R_0}$$

or

$$R = R_0 R_n$$

then the actual resistance level R is obtained by multiplying the normalized value by R_0. In the case of an inductance, its impedance is increased by multiplying it by R_0, but for a capacitance, its value of C must be *decreased* by R_0 in order to increase its impedance. Hence, when both frequency and resistance level are normalized, we have the equations

$$R = R_n R_0$$

$$L = \frac{L_n R_0}{\omega_c}$$

$$C = \frac{C_n}{\omega_c R_0}$$

EXAMPLE 16

Design a third-order Butterworth filter with a cut-off frequency $\omega_c = 10^6$ rad/s and a load impedance of 600 Ω.

Solution

The filter elements are designed using normalized values of $\omega = 1$ rad/s and a load impedance of 1 Ω.

Since $n = 3$, from Table 2 we obtain

$$F(s) = \frac{1}{s^3 + 2s^2 + 2s + 1}$$

For a ladder network, with $n = 3$, there are three reactive components L_1, L_3 and C_2 and the network is as shown in Fig. 49.

In this case, it can be shown that

$$F(s) = \frac{V_2(s)}{V_1(s)} = \frac{1}{(s^3 L_1 L_3 C_2/R) + s^2 L_1 C_2 + [s(L_1 + L_3)/R] + 1}$$

By equating terms, we obtain

$$\frac{L_1 L_3 C_2}{R} = 1 \quad \text{or} \quad L_1 L_3 C_2 = 1$$

$$L_1 C_2 = 2 \qquad L_1 C_2 = 2$$

$$\frac{L_1 + L_3}{R} = 2 \qquad L_1 + L_3 = 2$$

Hence
$$L_1 = \tfrac{3}{2}\,\text{H}$$
$$L_3 = \tfrac{1}{2}\,\text{H}$$
$$C_2 = \tfrac{4}{3}\,\text{F}$$

To obtain the *denormalized* values R'_0, L'_1, L'_3 and C'_2 we have

$$R'_0 = R_0 R_\text{n} = 600 \times 1 = 600\ \Omega$$

$$L'_1 = \frac{L_1 R_0}{\omega_\text{c}} = \frac{3}{2} \times \frac{600}{10^6} = 0\!\cdot\!9\ \text{mH}$$

$$L'_3 = \frac{L_3 R_0}{\omega_\text{c}} = \frac{1}{2} \times \frac{600}{10^6} = 0\!\cdot\!3\ \text{mH}$$

$$C'_2 = \frac{C_2}{\omega_\text{c} R_0} = \frac{4}{3 \times 10^6 \times 600} = 0\!\cdot\!0022\ \mu\text{F}$$

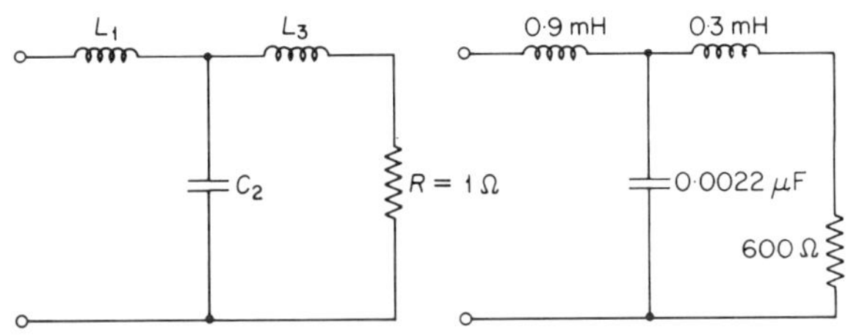

Fig. 49

6.4   Chebyshev filter[11-14]

The Chebyshev approximation of the ideal low-pass filter response is shown in Fig. 50 and given by

$$F_\text{n}(s) \cdot F_\text{n}(-s) = |F_\text{n}(s)|^2 = \frac{1}{1 + \varepsilon^2 C_\text{n}^2(\omega)}$$

where $\varepsilon < 1$ and $C_\text{n}(\omega)$ are the Chebyshev polynomials of order n.

Since the approximation tends to fluctuate between equal limits over the pass band, it is often called the 'equal-ripple' approximation.

The Chebyshev polynomials $C_\text{n}(\omega)$ are given by

$$C_\text{n}(\omega) = \cos(n \cos^{-1} \omega) \qquad \text{when} \quad 0 \leqslant \omega \leqslant 1$$

and
$$C_\text{n}(\omega) = \cosh(n \cosh^{-1} \omega) \quad \text{when} \quad \omega \geqslant 1$$

Typical values are shown in Table 3.

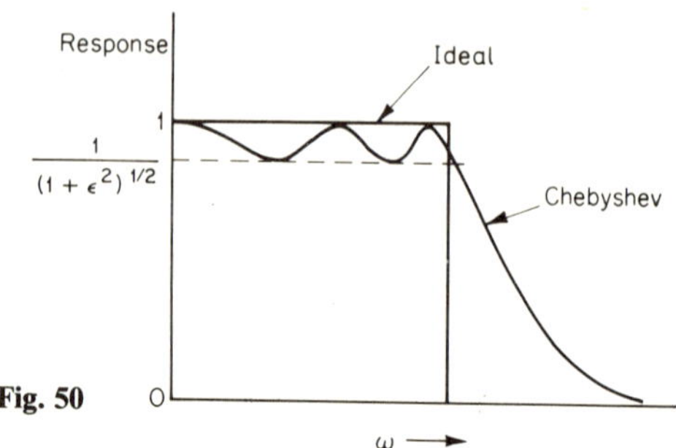

Fig. 50

Table 3

n	$C_n(\omega)$
1	ω
2	$2\omega^2 - 1$
3	$4\omega^3 - 3\omega$
4	$8\omega^4 - 8\omega^2 + 1$

The 'ripples' vary between 1 and $(1 + \varepsilon^2)^{-1/2}$ giving a *ripple height* of $1 - (1 + \varepsilon^2)^{-1/2}$ over the pass band. Well beyond the pass band, when $\omega > 1$, we have

$$F_n(s) = \frac{1}{\varepsilon C_n(\omega)}$$

or

$$20 \log_{10} F_n(s) = -20 \log_{10} \varepsilon C_n(\omega)$$

$$= -20 \log_{10} \varepsilon - 20 \log_{10} C_n(\omega)$$

For $\omega > 1$, $C_n(\omega) = 2^{n-1}(\omega^n)$

Hence

$$20 \log_{10} F_n(s) = -[20 \log_{10} \varepsilon + 6{\cdot}02(n - 1) + 20n \log_{10} \omega]$$

or

$$\text{db loss} = 20 \log_{10} \varepsilon + 6{\cdot}02(n - 1) + 20n \log_{10} \omega$$

At high frequencies, the fall-off is therefore $20n$ db/decade, the same as for the Butterworth filter. However, this approximation depends upon the two variables ε and n, the former defining the maximum permissible ripple, and the latter the required attenuation.

The poles of the filter are given by

$$1 + \varepsilon^2 C_n^2\left\{\frac{s}{j}\right\} = 0$$

or

$$C_n\left(\frac{s}{j}\right) = \pm j\,\frac{1}{e}$$

Now

$$C_n\left(\frac{s}{j}\right) = \cos\left[n \cos^{-1}\left(\frac{s}{j}\right)\right]$$

and it may be assumed that

$$\cos^{-1}\left(\frac{s}{j}\right) = \gamma + j\beta$$

Hence

$$\cos\left[n \cos^{-1}\left(\frac{s}{j}\right)\right] = \cos(n\gamma + nj\beta)$$

or

$$\cos(n\gamma + nj\beta) = \pm j\,\frac{1}{\varepsilon}$$

Equating real and imaginary parts leads to

$$\cos n\gamma \,.\, \cosh n\beta = 0$$

$$\sin n\gamma \,.\, \sinh n\beta = \pm\frac{1}{\varepsilon}$$

Since

$$\cosh n\beta \neq 0$$

We have

$$\cos n\gamma = 0$$

or

$$\gamma = \pm\frac{\pi}{2n}\,,\ \pm\frac{3\pi}{2n}\ \text{etc.}$$

Hence

$$\sin n\gamma = \pm 1$$

and

$$\sinh n\beta = \frac{1}{\varepsilon}$$

or

$$\beta = \frac{1}{n}\sinh^{-1}\frac{1}{\varepsilon}$$

Since γ and β are thus known, s is obtained from the equation

$$s = j\cos(\gamma + j\beta)$$

or

$$s = \sin\gamma \,.\, \sinh\beta + j\cos\gamma \,.\, \cosh\beta$$

As an example let $n = 2$ and $\varepsilon = 0\cdot 5$

Hence

$$\gamma = \pm\frac{\pi}{4}$$

$$\beta = \tfrac{1}{2}\sinh^{-1}\frac{1}{0\cdot 5} = 0\cdot 72$$

and

$$s = \sin\frac{\pi}{4} \,.\, \sinh 0\cdot 72 \pm j\cos\frac{\pi}{4} \,.\, \cosh 0\cdot 72$$

or

$$s_1 = 0\cdot 554 + j0\cdot 9$$

$$s_2 = 0\cdot 554 - j0\cdot 9$$

and

$$F(s) = \frac{1}{(s + s_1)(s + s_2)}$$

$$= \frac{1}{(s + 0\cdot 554 + j0\cdot 9)(s + 0\cdot 554 - j0\cdot 9)}$$

or

$$F(s) = \frac{1}{s^2 + 1\cdot 108s + 1\cdot 117}$$

EXAMPLE 17

Design a Chebyshev low-pass filter with a cut-off frequency of 10^4 rad/s and a design impedance of 600 Ω. The ripple in the pass-band must not exceed 0·5 db and the response after cut-off must be -60 db/decade.

Solution

The response after cut-off is $-20n$ db/decade and so $n = 3$, i.e. it must be a third-order filter.

The ripple amplitude in the pass band is given by

$$20\log(1 + \varepsilon^2)^{1/2} = 0\cdot 5$$

or

$$1 + \varepsilon^2 = 1\cdot 122$$

or

$$\varepsilon = 0\cdot 349$$

Assuming normalized values of $\omega = 1$, $R_0 = 1\,\Omega$, β is obtained from the equation

$$\beta = \frac{1}{n}\sinh^{-1}\frac{1}{\varepsilon}$$

$$= \frac{1}{3}\sinh^{-1}\frac{1}{0\cdot349}$$

or $$\beta = 0\cdot097$$

Also $$\sin n\gamma = \pm1$$

or $$\gamma = \pm\frac{\pi}{2n}\,,\,\pm\frac{3\pi}{2n}$$

Hence $$\gamma = +\frac{\pi}{6}\,,\,-\frac{\pi}{6}\,,\,+\frac{\pi}{2}$$

The plots are s_1, s_2, s_3 where

$$s_1 = j\cos(\gamma + j\beta)$$

$$= \sin\gamma\,.\,\sinh\beta + j\cos\gamma\,.\,\cosh\beta$$

$$= \sin\frac{\pi}{6}\,.\,\sinh 0\cdot097 + j\cos\frac{\pi}{6}\,.\,\cosh 0\cdot097$$

$$= 0\cdot5 \times 0\cdot1 + j0\cdot866 \times 0\cdot1$$

or $$s_1 = 0\cdot05 + j0\cdot0866$$

and $$s_2 = 0\cdot05 - j0\cdot0866 \quad \text{(conjugate pole)}$$

with $$s_3 = \sin\frac{\pi}{2}\,.\,\sinh 0\cdot097 = 1 \times 0\cdot1$$

or $$s_3 = 0\cdot1$$

and $$F(s) = \frac{1}{(s + 0\cdot05 + j0\cdot0866)(s + 0\cdot05 - j0\cdot0866)(s + 0\cdot1)}$$

$$= \frac{1}{s^3 + 0\cdot2s^2 + 0\cdot0425s + 0\cdot00325}$$

or $$F(s) = \frac{1}{333s^3 + 66\cdot6s^2 + 14\cdot2s + 1}$$

By equating coefficients as in Example 18 we obtain

$$\frac{L_1 L_3 C_2}{R} = 333 \quad \text{or} \quad L_3 = \frac{333}{66 \cdot 6} = 5 \text{ H}$$

$$L_1 C_2 = 66 \cdot 6 \qquad L_1 = 14 \cdot 2 - 5 = 9 \cdot 2 \text{ H}$$

$$L_1 + L_3 = 14 \cdot 2 \qquad C_2 = \frac{66 \cdot 6}{9 \cdot 2} = 7 \cdot 23 \text{ F}$$

Denormalizing to $\omega = 10^4$ rad/s and $R = 600 \ \Omega$ yields

$$R_1 = R_0 R_n = 600 \times 1 = 600 \ \Omega$$

$$L_1' = \frac{L_1 R_0}{\omega_c} = \frac{9 \cdot 2 \times 600}{10^4} = 0 \cdot 552 \text{ H}$$

$$L_3' = \frac{L_3 R_0}{\omega_c} = \frac{5 \times 600}{10^4} = 0 \cdot 3 \text{ H}$$

$$C_2' = \frac{C_2}{\omega_c R_0} = \frac{7 \cdot 23}{10^4 \times 600} = 1 \cdot 2 \ \mu\text{F}$$

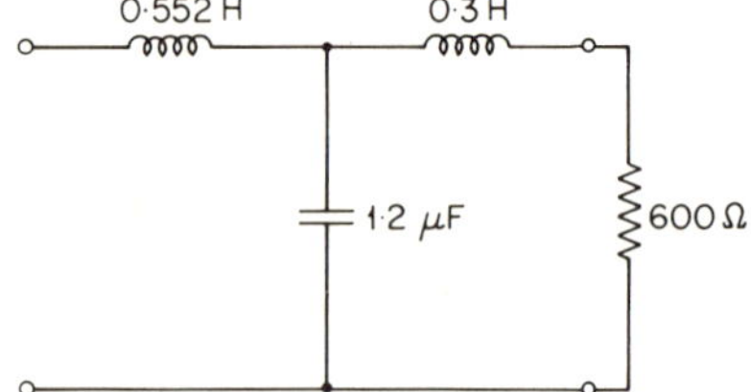

Fig. 51   The Filter

6.5   Frequency transformation

The previous design procedures have been applied essentially to low-pass filters. However, the high-pass filter can be designed simply by replacing inductors with capacitors, and capacitors with inductors, having the same reactance at the normalized frequency. The response characteristic is identical to that of the low-pass case and the frequency scale is inverted about the normalized frequency.

Band-pass filters are designed from low-pass filters by choosing the normalized point equal to the geometric mean of the bandwidth of the filter. Capacitors are then placed in series with inductors and with inductors in parallel with capacitors, such that the latter produce branch resonance at the centre frequency.

Band-stop filters may likewise be designed from the high-pass filter. The latter is scaled to a frequency which gives the rejection bandwidth required for the band-stop filter. Inductors and capacitors are then added to each branch in the same way as for the band-pass filter, with each branch resonating at the desired centre frequency which is the point of maximum attenuation.

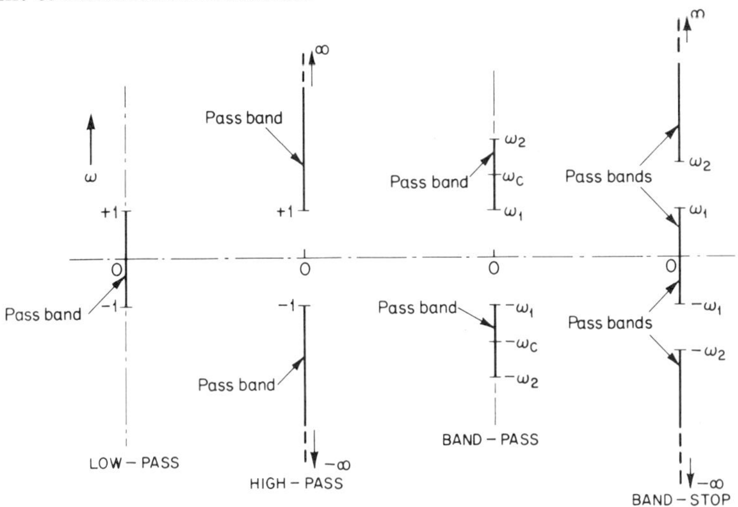

Fig. 52

(a)   High-pass filter

If s is any frequency variable and s_n is the normalized frequency for a *low-pass* filter with cut-off ω_c then

$$s_n = \frac{s}{\omega_c}$$

where $s = j\omega$.

For a high-pass filter, the frequency is inverted about the cut-off value such that

$$s_n = \frac{\omega_c}{s}$$

or $\qquad\qquad\qquad\qquad\qquad s = \dfrac{\omega_c}{s_n}$

Hence, over the pass-band, ω_n for the low-pass filter lies between $-1 \leqslant \omega_n \leqslant 1$, and so ω_n for the high-pass filter varies between ∞ and 1 or ∞ and -1.

To obtain the component values, let L_n and C_n be the component values for the low-pass filter and L_h, C_h those for the high-pass filter. Remembering that the components are interchanged with the same reactance, we have

$$s_n L_n = \frac{\omega_c L_n}{s} = \frac{1}{s C_h}$$

and

$$\frac{1}{s_n C_n} = \frac{s}{\omega_c C_n} = s L_h$$

Hence

$$L_h = \frac{1}{\omega_c C_n}$$

$$C_h = \frac{1}{\omega_c L_n}$$

The transformation is shown in Fig. 52.

(b)   Band-pass filter

The transformation from low pass to band pass is made by using the substitution

$$s_n = \frac{\omega_c}{\omega_2 - \omega_1}\left[\frac{s}{\omega_c} + \frac{\omega_c}{s}\right]$$

where $\omega_1 < \omega_2$ and ω_1, ω_2 are the two cut-off frequencies of the band-pass filter. Moreover, ω_c is the geometric mean frequency and is given by

$$\omega_c = \sqrt{\omega_1 \cdot \omega_2}$$

If the low-pass components are L_n and C_n, L_n transforms into a series L_s and C_s while C_n transforms into a parallel L_p and C_p. Their values are given by the equations

$$s_n L_n = \frac{\omega_c L_n}{\omega_2 - \omega_1}\left[\frac{s}{\omega_c} + \frac{\omega_c}{s}\right] = s L_s + \frac{1}{s C_s}$$

i.e. a series L_s and C_s.

Also

$$\frac{1}{s_n C_n} = \frac{\dfrac{1}{C_n}\left[\dfrac{\omega_2 - \omega_1}{\omega_c}\right]}{\left(\dfrac{s}{\omega_c} + \dfrac{\omega_c}{s}\right)}$$

$$= \frac{\dfrac{(\omega_2 - \omega_1)^2}{\omega_c^2 C_n^2}}{\dfrac{s(\omega_2 - \omega_1)}{\omega_c C_n} + \dfrac{(\omega_2 - \omega_1)}{s C_n}}$$

which is equivalent to

$$\frac{sL_p \times \dfrac{1}{sC_p}}{sL_p + \dfrac{1}{sC_p}}$$

i.e. a parallel L_p and C_p.

Hence we obtain by equating terms

$$L_s = \frac{L_n}{(\omega_2 - \omega_1)} \qquad C_s = \frac{\omega_2 - \omega_1}{\omega_c^2 L_n}$$

$$L_p = \frac{(\omega_2 - \omega_1)}{\omega_c^2 C_n} \qquad C_p = \frac{C_n}{(\omega_2 - \omega_1)}$$

The transformation is shown in Fig. 52.

(c) Band-stop filter

The transformation is obtained by modifying the expression for s_n as used in the band-pass case to give

$$s_n = \frac{\omega_2 - \omega_1}{\omega_c\left(\dfrac{s}{\omega_c} + \dfrac{\omega_c}{s}\right)}$$

In this case, the series L_n transforms into a parallel L_p, C_p and the shunt C_n transforms into a series L_s, C_s. They are given by the equations

$$s_n L_n = \frac{L_n(\omega_2 - \omega_1)}{\omega_c\left(\dfrac{s}{\omega_c} + \dfrac{\omega_c}{s}\right)} = \frac{\dfrac{L_n^2}{\omega_c^2}(\omega_2 - \omega_1)^2}{\dfrac{sL_n(\omega_2 - \omega_1)}{\omega_c^2} + \dfrac{1}{s}L_n(\omega_2 - \omega_1)}$$

which is equivalent to

$$\frac{sL_p \times \dfrac{1}{sC_p}}{sL_p + \dfrac{1}{sC_p}}$$

i.e. a parallel L_p and C_p.
  Also

$$\frac{1}{s_n C_n} = \frac{1}{C_n}\left[\frac{\omega_c}{\omega_2 - \omega_1}\right]\left(\frac{s}{\omega_c} + \frac{\omega_c}{s}\right) = \frac{s}{(\omega_2 - \omega_1)C_n} + \frac{1}{s}\frac{\omega_c^2}{(\omega_2 - \omega_1)C_n}$$

which is equivalent to

$$sL_s + \frac{1}{sC_s}$$

i.e. a series L_s and C_s.
  Hence we obtain by equating terms

$$L_p = \frac{L_n(\omega_2 - \omega_1)}{\omega_c^2} \qquad C_p = \frac{1}{(\omega_2 - \omega_1)L_n}$$

$$L_s = \frac{1}{(\omega_2 - \omega_1)C_n} \qquad C_s = \frac{(\omega_2 - \omega_1)C_n}{\omega_c^2}$$

The transformation is shown in Fig. 52.

EXAMPLE 18

Design a band-pass filter from the normalized low-pass structure shown below. The centre frequency is 4×10^4 rad/s. Determine the component values for a load impedance of 600 Ω.

Solution

Using impedance denormalization of the low-pass filter only yields, for a load impedance of 600 Ω,

$$R_0' = R_0 R_n = 600 \times 1 = 600 \ \Omega$$

$$L_n' = L_n R_0 = \tfrac{2}{3} \times 600 = 400 \ \text{H}$$

$$C_{n_1}' = \frac{C_{n_1}}{R_0} = \frac{1}{4 \times 600} = \frac{1250}{3} \ \mu\text{F}$$

$$C_{n_2}' = \frac{C_{n_2}}{R_0} = \frac{3}{4 \times 600} = 1250 \ \mu\text{F}$$

For a band-pass transformation, the inductance becomes a series L_s and C_s given by

$$L_s = \frac{L'_n}{\omega_2 - \omega_1} = \frac{400}{2 \times 10^4} = 20 \text{ mH}$$

$$C_s = \frac{\omega_2 - \omega_1}{\omega_c^2 L'_n} = \frac{2 \times 10^4}{(4 \times 10^4)^2 \times 400} = 0.312 \ \mu\text{F}$$

The capacitances C'_{n_1} and C'_{n_2} transform into parallel L_p and C_p given by

$$L_{p_1} = \frac{\omega_2 - \omega_1}{\omega_c^2 . C'_{n_1}} = \frac{2 \times 10^4 \times 3 \times 10^6}{(4 \times 10^4)^2 \times 1250} = 0.03 \text{ H}$$

$$C_{p_1} = \frac{C'_{n_1}}{\omega_2 - \omega_1} = \frac{1250 \times 10^{-6}}{3 \times 2 \times 10^4} = 0.0208 \ \mu\text{F}$$

$$L_{p_2} = \frac{\omega_2 - \omega_1}{\omega_c^2 . C'_{n_2}} = \frac{2 \times 10^4 \times 10^6}{(4 \times 10^4)^2 \times 1250} = 0.09 \text{ H}$$

$$C_{p_2} = \frac{C'_{n_2}}{\omega_2 - \omega_1} = \frac{1250 \times 10^{-6}}{2 \times 10^4} = 0.06 \ \mu\text{F}$$

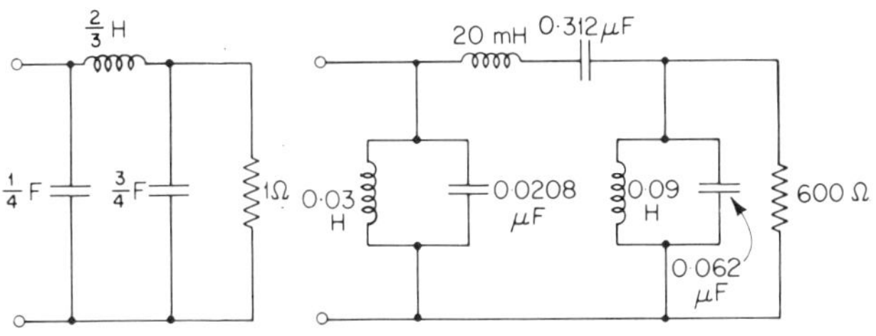

Fig. 53

6.6   Active filters[15-23]

In order to achieve the usual transfer functions attainable with passive *LC* filters, the design of *RC* filters requires the use of some active device to achieve similar results. Two important active devices used in the design of *RC* active filters is the negative-impedance converter (NIC)[15-17] and the gyrator.[20-21]

(a) Negative-impedance converter

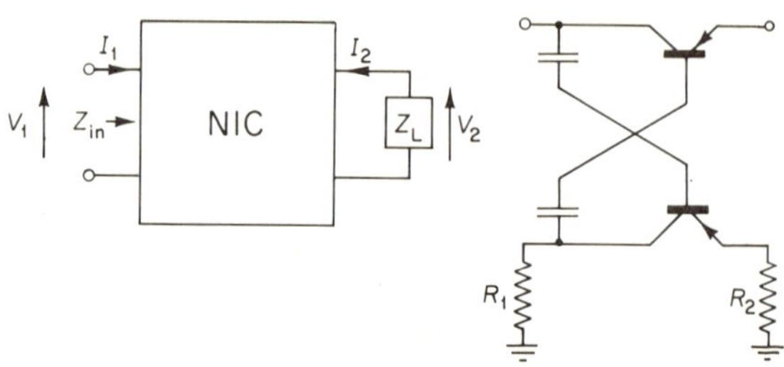

Fig. 54

The ideal negative-impedance converter is a two-port device whose input impedance is equal to the negative of the load impedance, i.e. $Z_{in} = -Z_L$ where Z_{in} is the input impedance and Z_L is the load impedance (Fig. 54).

Using g-parameters, we obtain the equations

$$I_1 = g_{11}V_1 + g_{12}I_2$$
$$V_2 = g_{21}V_1 + g_{22}I_2$$

with
$$Y_{in} = \frac{I_1}{V_1} = g_{11} - \frac{g_{12} \cdot g_{21}}{g_{22} + Z_L}$$

If
$$g_{11} = g_{22} = 0 \quad \text{and} \quad g_{12} \cdot g_{21} = 1$$

then
$$Y_{in} = -\frac{1}{Z_1}$$

or
$$Z_{in} = -Z_L$$

There are two types of NIC, the voltage-inversion NIC (VNIC) and the current-inversion NIC (INIC). The equivalent circuits are shown below.

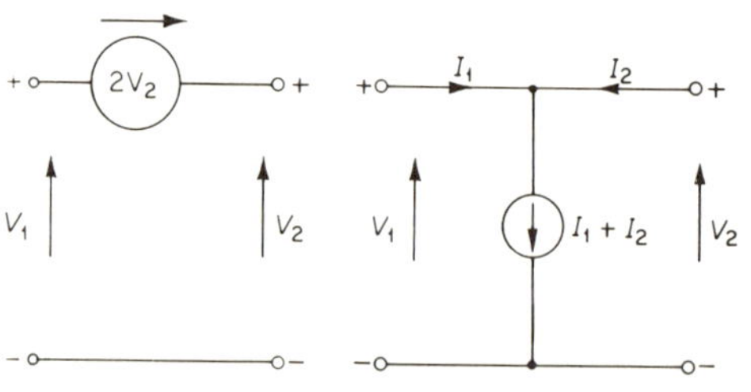

Fig. 55

An example of a VNIC is the circuit due to Linvill,[15] while one of an INIC is given by Yanagisawa.[17] A practical arrangement for synthesizing an *RC* active filter of the type due to Linvill is shown in Fig. 56.

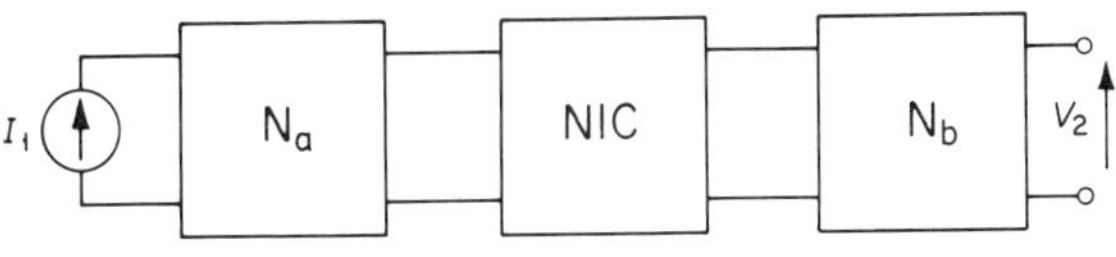

Fig. 56

In this method, two passive networks N_a and N_b are connected in cascade with an NIC. If Z_{21a}, Z_{21b} are the open-circuit transfer impedances and Z_{22a}, Z_{11b} are the driving-point impedances of the networks respectively, then the overall transfer impedance is given by

$$Z_{21} = \frac{Z_{21a} \cdot Z_{21b}}{Z_{22a} - Z_{11b}}$$

The poles of Z_{21} which determine the natural frequencies of the whole arrangement are given by $Z_{22a} - Z_{11b} = 0$. The zeros of transmission correspond to $Z_{21a} \cdot Z_{21b} = 0$ and are only dependent on the passive networks N_a and N_b, which may be suitably chosen. An example of a low-pass third-order Butterworth filter arrangement is shown in Fig. 57.

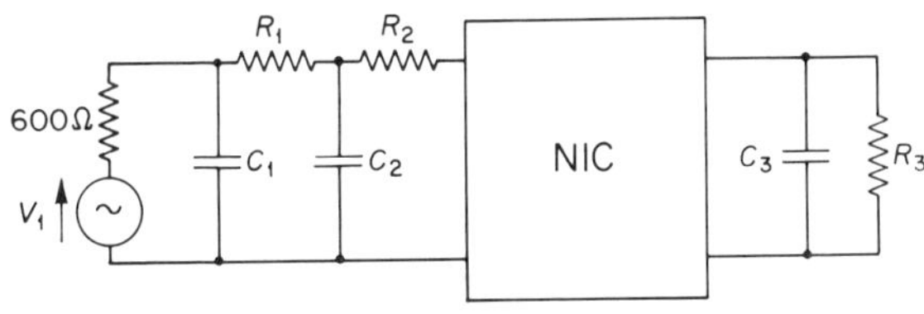

Fig. 57

(b)   The gyrator[20,21]

An ideal gyrator is a non-reciprocal two-port device whose input impedance is the reciprocal of the load impedance, i.e. $Z_{in} = R^2/Z_L$ where R is the gyration resistance and Z_L is the load impedance.

Using y-parameters, we obtain the equations:

$$I_1 = y_{11}V_1 + y_{12}V_2$$
$$I_2 = y_{21}V_1 + y_{22}V_2$$

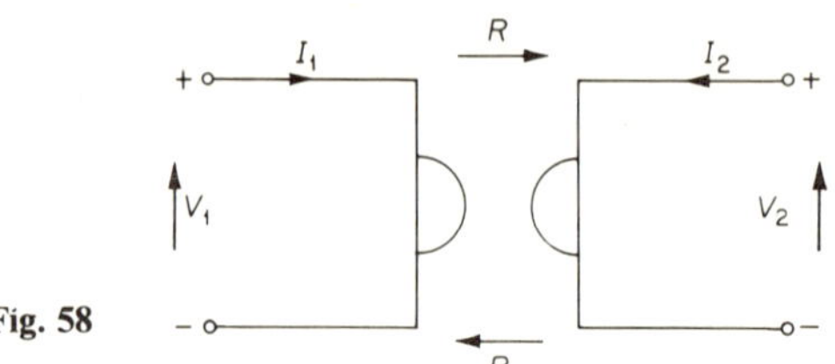

Fig. 58

or
$$Y_{in} = \frac{I_1}{V_1} = y_{11} - \frac{y_{12} \cdot y_{21}}{y_{22} + Y_L}$$

where Y_L is the load admittance.

If $y_{11} = y_{22} = 0$ and $y_{12} \cdot y_{21} = -g^2$ (passive gyrator) where $g = 1/R$ is the gyration conductance, then

$$Y_{in} = \frac{g^2}{Y_L}$$

or
$$Z_{in} = \frac{R^2}{Z_L}$$

If $y_{12} = g$ and $y_{21} = -g$ then

$$I_1 = gV_2 \qquad \text{or} \qquad \begin{bmatrix} I_1 \\ I_2 \end{bmatrix} = \begin{bmatrix} 0 & g \\ -g & 0 \end{bmatrix}\begin{bmatrix} V_1 \\ V_2 \end{bmatrix}$$
$$I_2 = -gV_1$$

Since I_1 is proportional to V_2 while I_2 is proportional to $-V_1$, the device is non-reciprocal. Moreover, the admittance matrix can be written as the sum of two admittances in parallel, viz.,

$$\begin{bmatrix} 0 & g \\ -g & 0 \end{bmatrix} = \begin{bmatrix} 0 & 0 \\ -g & 0 \end{bmatrix} + \begin{bmatrix} 0 & g \\ 0 & 0 \end{bmatrix}$$

Hence, the gyrator can be conceived as two voltage-controlled current sources in parallel.

The gyrator was proposed by Tellegen.[20] Because of its inverting property, a gyrator when terminated by a capacitance behaves as an inductance. Gyrators can also behave as isolators and circulators when suitably terminated. An *RC* active filter can be constructed using a gyrator in much the same way as an NIC; an arrangement is shown in Fig. 59.

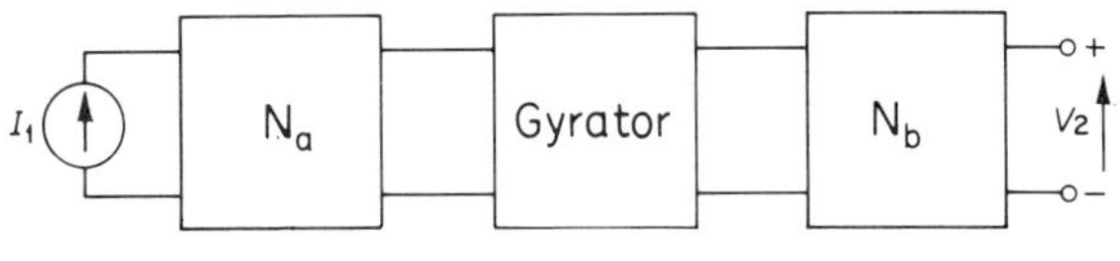

Fig. 59

If y_{21a}, y_{22a} are the driving-point admittances and Z_{11a}, Z_{21b} are the driving-point impedances, the overall open-circuit voltage transfer function is given by

$$Z_{21} = \frac{-gy_{21a} \cdot Z_{21b}}{y_{22a} + g^2 Z_{11b}}$$

where g is the gyration conductance.

By assuming that the poles of y_{21a} and y_{22a} are the same, as also the poles of Z_{11b} and Z_{21b}, an analysis due to Calahan[22] shows that Z_{21} can be realized.

(c) Sensitivity[22,23]

One of the most important problems associated with *RC* active filters is their sensitivity to changes in component values and conversion factors. This arises because the poles of the transfer impedance Z_{21} depends on the difference or sum of terms in the denominator, each of which may change with component changes. A study of the stability of such circuits has held the attention of many researchers in recent years.

Problems

1 A tuned circuit consists of pure resistance, pure inductance and pure capacitance, all connected in parallel. Show that the ratio y, of the circuit admittance to admittance at resonance is given by

$$y = 1 + jx$$

where x is a function of the frequency f, the resonant Q-factor Q_0 and the resonant frequency f_0.

For small frequency deviations Δf from the resonant value, show that

$$x = \frac{2Q_0\,\Delta f}{f_0}$$

Hence, determine the Q-factor of the tuned circuits required for the anode loads of a 5-stage amplifier, tuned to 90 MHz, if the overall bandwidth between the 3 dB points is to be 3 MHz. The stages are all similar, each using a pentode valve with a single circuit, tuned to 90 MHz, as the anode load.

L.U.B.Sc(Eng) Tels. Pt. 3., 1966

2 A voltage is induced in a series circuit consisting of an inductance L, an effective resistance R of 12 Ω and a capacitance C. The Q-factor of the coil is 100 and the circuit is tuned to resonance at 1 MHz. Determine the ratio of the voltage across the capacitor at resonance to that across it at 10 kHz off resonance and the value of C.

3 Design a series Foster network to have a pole at $\omega = 10^6$ rad/s and a zero at $\omega = 2 \times 10^6$ rad/s. The impedance of the network is j250 Ω at $\omega = 0{\cdot}5 \times 10^6$ rad/s.

4 A reactance function has an internal zero at $\omega = 2 \times 10^6$ rad/s and an internal pole at $\omega = 10^7$ rad/s. The driving point impedance is j200 Ω at $\omega = 8 \times 10^6$ rad/s. Sketch the reactance curve and determine the first Foster network.

5 Synthesize the first Cauer network for the driving-point impedance given by

$$Z(s) = \frac{2s^4 + 8s^2 + 6}{s^3 + 2s}$$

6 Define the terms image and iterative impedance as applied to two-port networks. Calculate the iterative impedances for the following network.

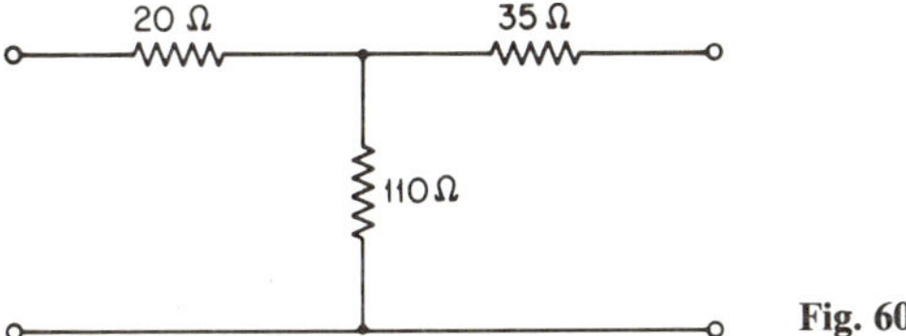

Fig. 60

What is the insertion loss of the network when it is inserted between its iterative impedances?   I.E.E. Advanced Elect. Eng. Pt. 3, 1961

7 Determine the passive T network shown below, which is equivalent from the point of view of transmission between terminals pairs 1, 2 and 3, 4, to the passive lattice network shown alongside.

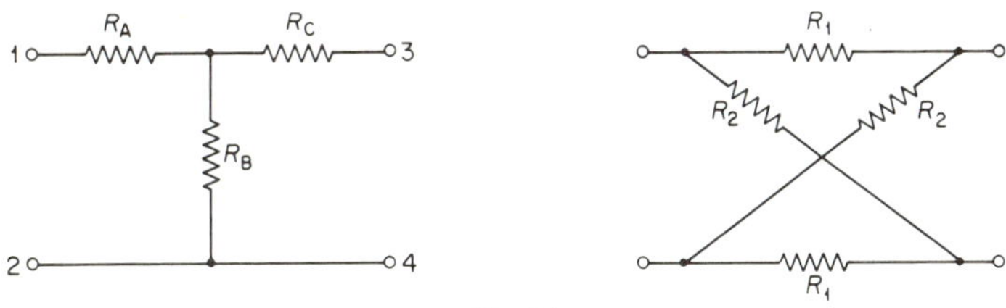

Fig. 61

I.E.E. Line Communication Pt. 3, 1961

8 Distinguish between iterative impedance and image impedance of a network and give examples of the application of these impedances in circuit design. Calculate the iterative and image impedances at the terminals 1, 2 and 3, 4 of the network below.

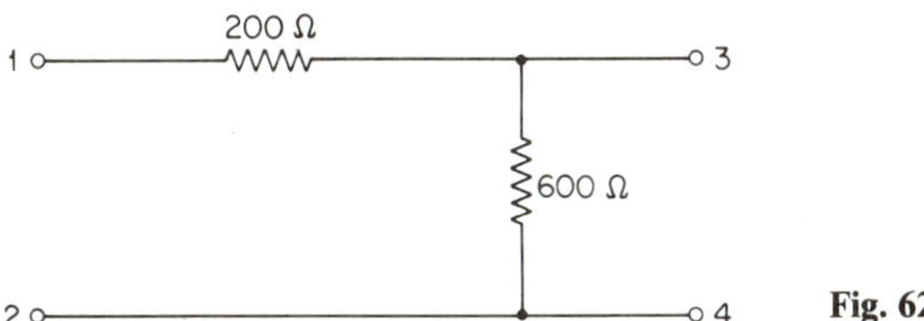

Fig. 62

9 A low-pass filter is composed of π sections having series arms of 63·6 mH inductance and shunt arms each of 0·088 μF capacitance.

Determine, either by deriving the necessary formulae or by the direct application of simple circuit theory, (a) the cut-off frequency and (b) the attenuation per section at a frequency of 5 kHz. Assume that the filter is terminated in its characteristic impedance and that the elements are loss free.   L.U.B.Sc(Eng) Tels., 1948

10 Design a prototype constant-k, T-section, high-pass filter to have a cut-off frequency of $10{,}000/2\pi$ Hz and characteristic impedance of 500 Ω at a very high frequency.

   The equations

$$\cosh \gamma = 1 + \frac{Z_1}{2Z_2}$$

and

$$Z_k = \left(Z_1 Z_2 + \frac{Z_1^2}{4}\right)^{1/2}$$

may be assumed, where γ = the propagation constant, Z_1 is the total series impedance, Z_2 is the shunt impedance and Z_k is the characteristic impedance of a T section. Explain your solution.

   Discuss the imperfections of the prototype and describe briefly how they may be overcome.   L.U.B.Sc(Eng) Tels. Pt. 3, 1960

11 The elements of a band-pass filter consist of an inductance and capacitance in series and another capacitance in shunt. Deduce the expressions for the cut-off frequencies.

   If the coils available have an inductance of 10 mH, determine the capacitance required to give cut-off frequencies of $20{,}000/2$ Hz and $40{,}000/2$ Hz.

   How will resistance and losses modify the shape of the attenuation/frequency curve?   L.U.B.Sc(Eng) 1948

12 An m-derived π-type low-pass filter section is to satisfy the following requirements:

   Cut-off frequency 3500 Hz, frequency of infinite attenuation 4000 Hz, nominal characteristic impedance 600 Ω. Calculate the values of the elements required.

13 Design an m-derived π-section, high-pass filter with a nominal impedance of 600 Ω, a cut-off frequency of 10 kHz and a frequency of infinite attenuation of 8 kHz.

14 Design a Chebyshev filter with the specification $\varepsilon = 0{\cdot}67$, $\omega_c = 1$ rad/s and $R_L = 1\ \Omega$, for the circuit below.

   Hence obtain the circuit components if $\omega_c = 10{,}000$ rad/s and $R_L = 1{\cdot}0$ kΩ.

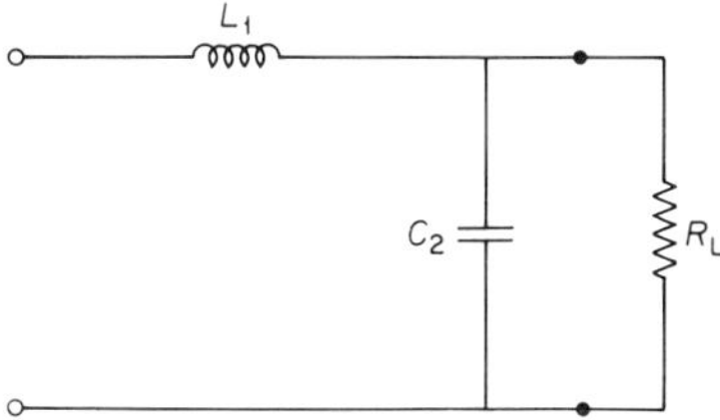

Fig. 63

15 For the normalized low-pass filter shown below, design a band-pass filter with a centre frequency of $\omega_c = 50,000$ rad/s. The filter bandwidth is 25,000 rad/s. Hence, determine the circuit elements for $R_L = 1$ kΩ.

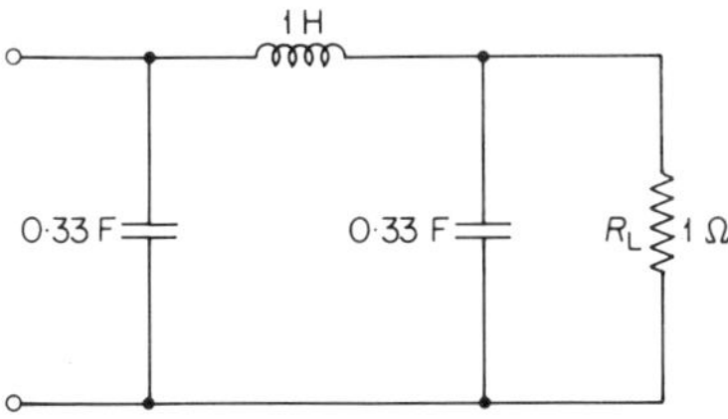

**Fig. 64

Answers

1   11·58
2   2·84
    133 pf
3   $L_0 = 0{\cdot}1$ mH, $L_1 = 0{\cdot}3$ mH, $C_1 = 0{\cdot}0033\ \mu$F
4   $C_0 = 0{\cdot}026\ \mu$F
    $L_2 = 9{\cdot}1\ \mu$H
    $C_2 = 0{\cdot}0011\ \mu$F
5   Series arms: 2 H and 8 H
    Shunt arms: $\frac{1}{4}$ F and $\frac{1}{12}$ F
6   75 Ω, 90 Ω, 6 dB
7   $R_A = R_C = R_1$, $R_B = (R_2 - R_1)/2$
8   461 Ω and 261 Ω
    400 Ω and 300 Ω
9   3 kHz, 19·1 dB
10   $0{\cdot}2\ \mu$F, 25 mH
11   $0{\cdot}25\ \mu$F, $0{\cdot}166\ \mu$F
12   Series arm: 26·4 mH in parallel with $0{\cdot}0597\ \mu$F
     Shunt arm: $0{\cdot}0366\ \mu$F
13   Series arm: 17·9 mH in parallel with $0{\cdot}022\ \mu$F
     Shunt arm: 15·9 mH
14   0·1 H
     $0{\cdot}112\ \mu$F
15   Series arm: 40 mH in series with $0{\cdot}01\ \mu$F
     Shunt arms: 30·3 mH in parallel with $0{\cdot}013\ \mu$F
              10 mH in parallel with $0{\cdot}04\ \mu$F

References

1 FOSTER, R. M. A reactance theorem. *Bell System Technical Journal*, 3-2 (April 1924), 259–267.

2 EVERITT, W. L. and ANNER, G. E. *Communication Engineering*. Third edition. McGraw-Hill (1956).

3 CAUER, W. The realisation of impedances with prescribed frequency dependence. *Archiv fur Electrotechnic*, 15 (1926), 355–358.

4 CAUER, W. *Synthesis of Linear Communication Networks*. McGraw-Hill (1958).

5 ZOBEL, O. J. Theory and design of uniform and composite electric wave filters. *Bell System Technical Journal*, 2-1 (January 1923), 1–46.

6 ZOBEL, O. J. Transmission characteristics of electric wave-filters. *Bell System Technical Journal*, 3-4 (October 1924), 567–620.

7 VAN VALKENBURG, M. E. *Introduction to Modern Network Synthesis*, John Wiley & Sons Inc. New York (1960).

8 BUTTERWORTH, S. On the theory of filter amplifiers. *Wireless Engineer*, 7 (October 1930), 536–541.

9 DARLINGTON, S. Synthesis of reactance 4-poles which produce prescribed insertion loss characteristics. *Journal of Mathematics and Physics*, 18 (September 1939), 257–353.

10 LANDON, V. D. Cascade amplifiers with maximal flatness. *RCA Review*, 5 (1941), 34.

11 CHEBYSHEV, P. L. Théorie de parallelogrammes. *Oeuvres*. Volume 1.

12 STORER, J. E. *Passive Network Synthesis*. McGraw-Hill (1957). Page 293.

13 GUILLEMIN, E. A. *Synthesis of Passive Networks*. Wiley: Chapman & Hall (1957). Pages 107–134.

14 TUTTLE, D. F. *Network Synthesis*. Volume 1. Wiley: Chapman & Hall (1958).

15 LINVILL, J. G. RC active filters. *Proceedings Institute Radio Engineers*, 42 (March 1954), 555–564.

16 HOROWITZ, I. M. Optimisation of negative-impedance conversion methods of active RC synthesis. *Transactions Institute of Radio Engineers*, **CT-6,** (September 1959), 296–303.

17 YANAGISAWA, T. RC active networks using current inversion type negative impedance converters. *Transactions Institute of Radio Engineers*, **CT-4,** (September 1957), 140–144.

18 SALLEN, R. P. and KEY, E. L. A practical method of designing RC active filters. *Transactions Institute of Radio Engineers*, **CT-2** (1955), 74–85.

19 PIERCEY, R. N. Synthesis of active RC filter networks. *A.T.E. Journal*, **21**-2 (April 1965), 61–75.

20 TELLEGEN, B. D. The gyrator, a new electric network element. *Philips Research Report*, 3-2 (April 1948), 81–101.

21 SHENOI, B. A. Practical realisation of a gyrator circuit and RC-gyrator filters. *Transactions Institute of Radio Engineers*, **CT-12** (September 1965), 374–380.

22 CALAHAN, D. A. Sensitivity minimisation in active RC synthesis. *Transactions Institute of Radio Engineers*, **CT-9** (March 1962), 38–42.

23 WOODWARD, J. and NEWCOMB, R. W. Sensitivity improvement of inductorless filters. *Electronic Letters*, 2–9 (September 1966), 349–350.

Appendices

Appendix A: Band-pass filter
This passes all frequencies *within* a band defined by two cut-off values f_1 and f_2. Such filters are used very extensively in radio circuits, viz. the IF transformer.

Band passing may be achieved by cascading a low-pass filter and a high-pass filter, with the cut-off of the former greater than that of the latter. In practice, however, it is more economical and simpler to use the configuration shown in Fig. 65.

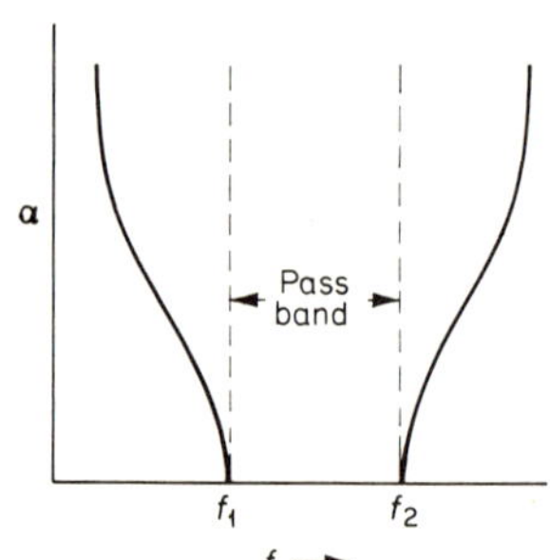

Fig. 65

Here

$$Z_1 = j\left[\frac{\omega^2 L_1 C_1 - 1}{\omega C_1}\right]$$

$$Z_2 = \frac{\omega L_2}{j(\omega^2 L_2 C_2 - 1)}$$

and

$$Z_1 Z_2 = \left(\frac{\omega^2 L_1 C_1 - 1}{\omega^2 L_2 C_2 - 1}\right)\left(\frac{L_2}{C_1}\right)$$

The filter is a constant–k type if $Z_1 Z_2 = k^2$ and this is true if $L_1 C_1 = L_2 C_2$.

Then

$$Z_1 Z_2 = \frac{L_1}{C_2} = \frac{L_2}{C_1} = k^2$$

The attenuation curves for the filter are shown in Fig. 65. The pass-band extends from f_1 to f_2.

At the cut-off frequencies, $Z_{0T} = 0$

or
$$\frac{Z_1}{4} + Z_2 = 0$$

with
$$Z_1^2 = -4Z_1 Z_2 = -4R_0^2 \quad \text{where} \quad R_0^2 = Z_1 Z_2$$

and
$$Z_1 = \pm j2R_0$$

Hence
$$j\left[\frac{\omega^2 L_1 C_1 - 1}{\omega C_1}\right] = \pm j2R_0$$

or
$$\omega^2 L_1 C_1 - 1 = \pm j\omega 2R_0 C_1$$

The solutions of this equation are

$$\omega_1 = \frac{-2R_0 C_1 + \sqrt{4R_0^2 C_1^2 + 4L_1 C_1}}{2L_1 C_1}$$

$$\omega_2 = \frac{2R_0 C_1 + \sqrt{4R_0^2 C_1^2 + 4L_1 C_1}}{2L_1 C_1}$$

giving
$$\omega_1 \cdot \omega_2 = \frac{1}{L_1 C_1} = \omega_0$$

where ω_0 is the geometric mean frequency of the pass-band and

$$\omega_2 - \omega_1 = \frac{2R_0}{L_1}$$

Hence
$$L_1 = \frac{R_0}{\pi(f_2 - f_1)}$$

$$C_1 = \frac{1}{\omega_1 \cdot \omega_2 \cdot L} = \frac{f_2 - f_1}{4\pi f_1 \cdot f_2 \cdot R_0}$$

Now
$$R_0^2 = \frac{L_1}{C_2} = \frac{L_2}{C_1}$$

Hence
$$L_2 = R_0^2 C_1 = \frac{R_0(f_2 - f_1)}{4\pi f_1 \cdot f_2}$$

and
$$C_2 = \frac{L_1}{R_0^2} = \frac{1}{\pi R_0(f_2 - f_1)}$$

EXAMPLE 19

Design a T-section band-pass filter to pass all frequencies between 300 Hz and 1000 Hz. The load impedance is 600 Ω

Here $\qquad f_1 = 400, \qquad f_2 = 1000, \qquad R_0 = 600$

Hence $\qquad\qquad\qquad f_2 - f_1 = 600$

giving $\quad L_1 = \dfrac{R_0}{\pi(f_2 - f_1)} = \dfrac{600}{\pi \cdot 600} = 0\cdot318 \text{ H}$

$$C_1 = \frac{f_2 - f_1}{4\pi f_1 \cdot f_2 \cdot R_0} = \frac{600}{4\pi \cdot 400 \cdot 1000 \cdot 600} = 0\cdot199 \ \mu\text{F}$$

$$L_2 = \frac{R_0(f_2 - f_1)}{4\pi f_1 \cdot f_2} = \frac{600 \cdot 600}{4\pi \cdot 400 \cdot 1000}$$

$$= 0\cdot0715 \text{ H}$$

$$C_2 = \frac{1}{\pi R_0(f_2 - f_1)} = \frac{1}{\pi \cdot 600 \cdot 600}$$

$$= 0\cdot0885 \ \mu\text{F}$$

Appendix B: Band-stop filter

It is sometimes useful to pass frequencies on *either* side of a given band (the stop band). This is possible, by placing a low-pass and a high-pass filter in parallel, with the correct cut-off frequencies f_1 and f_2 respectively. However, it is more useful to use a single-section structure as shown in Fig. 66.

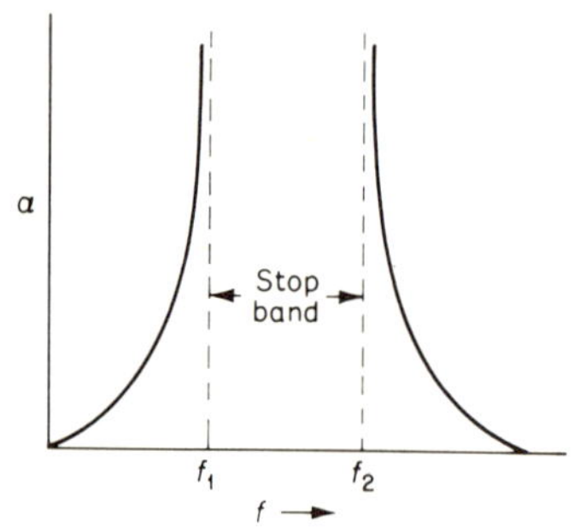

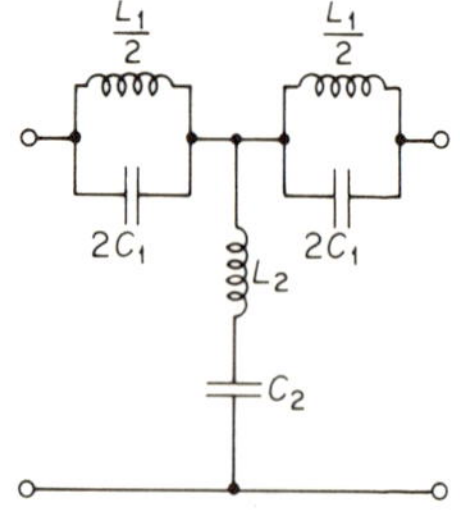

Fig. 66

Here
$$Z_2 = j\left(\omega L_2 - \frac{1}{\omega C_2}\right) = -j\left(\frac{1}{\omega C_2} - \omega L_2\right)$$

and it is usually arranged that $L_1 C_1 = L_2 C_2$ if it is of the constant–k type

This gives

$$R_0^2 = \frac{L_1}{C_2} = \frac{L_2}{C_1}$$

as in the case of the band-pass filter. The attenuation curves are shown in Fig. 66.

At the cut-off frequencies f_1 and f_2 we have

$$\frac{Z_1}{4} + Z_2 = 0$$

with
$$Z_1 = -4Z_2$$

or
$$Z_1 Z_2 = -4Z_2^2 = R_0^2$$

giving
$$Z_2 = \pm\frac{jR_0}{2}$$

i.e.
$$-j\left(\frac{1}{\omega_1 C_2} - \omega_1 L_2\right) = -\frac{jR_0}{2}$$

or
$$1 - \omega_1^2 L_2 C_2 = \frac{\omega_1 R_0 C_2}{2}$$

Now if
$$\omega_1 . \omega_2 = \omega_0^2 = \frac{1}{L_2 C_2}$$

then
$$1 - \left(\frac{f_1}{f_0}\right)^2 = \pi f_1 R_0 C_2$$

and
$$C_2 = \frac{f_2 - f_1}{\pi R_0 f_1 . f_2}$$

$$L_2 = \frac{R_0}{4\pi(f_2 - f_1)}$$

$$L_1 = \frac{R_0(f_2 - f_1)}{\pi f_1 . f_2}$$

and
$$C_1 = \frac{1}{4\pi R_0(f_2 - f_1)}$$

EXAMPLE 20

Design a T-section band-stop filter to stop all frequencies between 400 Hz and 1000 Hz. The load impedance is 600 Ω.

Here
$$f_1 = 400, \qquad f_2 = 1000$$
$$f_2 - f_1 = 600, \qquad R_0 = 600$$

Hence
$$L_1 = \frac{R_0(f_2 - f_1)}{\pi f_1 \cdot f_2} = \frac{600 \cdot 600}{\pi \cdot 400 \cdot 1000}$$

$$= 0 \cdot 286 \text{ H}$$

$$C_1 = \frac{1}{4\pi R_0(f_2 - f_1)} = \frac{1}{4\pi \cdot 600 \cdot 600}$$

$$= 0 \cdot 221 \ \mu\text{F}$$

$$L_2 = \frac{R_0}{4\pi(f_2 - f_1)} = \frac{600}{4\pi \cdot 600}$$

$$= 0 \cdot 0795 \text{ H}$$

$$C_2 = \frac{f_2 - f_1}{\pi R_0 f_1 \cdot f_2} = \frac{600}{\pi \cdot 600 \cdot 400 \cdot 1000}$$

$$= 0 \cdot 795 \ \mu\text{F}$$

Appendix C: Network parameters

The analysis of a two-port network can be undertaken through the use of various parameters. Each parameter may be determined by an open-circuit or short-circuit measurement on the network, where the positive direction of voltages and currents are shown in Fig. 67.

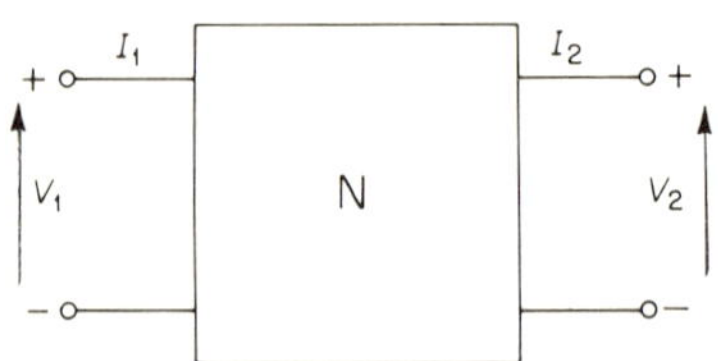

Fig. 67

(a) *A, B, C, D parameters*

The input voltage and current are given in terms of the output voltage and current. The load current I_2 is negative when it flows outwards into a load.

$$V_1 = AV_2 - BI_2$$
$$I_1 = CV_2 - DI_2$$

where

$$A = V_1/V_2 \quad \text{with} \quad I_2 = 0$$
$$B = -V_1/I_2 \quad \text{with} \quad V_2 = 0$$
$$C = I_1/V_2 \quad \text{with} \quad I_2 = 0$$
$$D = -I_1/I_2 \quad \text{with} \quad V_1 = 0$$

For any network it may be shown that $AD - BC = 1$ since A and D are pure numbers, while B is an impedance and C is an admittance.

(b) *z parameters*

The input and output voltages are expressed in terms of the input and output currents. This leads to the impedance parameters of the network.

$$V_1 = z_{11}I_1 + z_{12}I_2$$
$$V_2 = z_{21}I_1 + z_{22}I_2$$

where

$$z_{11} = V_1/I_1 \quad \text{with} \quad I_2 = 0$$
$$z_{12} = V_1/I_2 \quad \text{with} \quad I_1 = 0$$
$$z_{21} = V_2/I_1 \quad \text{with} \quad I_2 = 0$$
$$z_{22} = V_2/I_2 \quad \text{with} \quad I_1 = 0$$

(c) *y parameters*

The input and output currents are expressed in terms of the input and output voltages. This leads to the admittance parameters of the network.

$$I_1 = y_{11}V_1 + y_{12}V_2$$
$$I_2 = y_{21}V_1 + y_{22}V_2$$

where

$$y_{11} = I_1/V_1 \quad \text{with} \quad V_2 = 0$$
$$y_{12} = I_1/V_2 \quad \text{with} \quad V_1 = 0$$
$$y_{21} = I_2/V_1 \quad \text{with} \quad V_2 = 0$$
$$y_{22} = I_2/V_2 \quad \text{with} \quad V_1 = 0$$

(d) *h parameters*

The input voltage and output current are related to the input current and output voltage. These 'hybrid' parameters are used extensively with transistor circuits and are so called, because they may represent an impedance, admittance or a pure number.

$$V_1 = h_{11}I_1 + h_{12}V_2$$
$$I_2 = h_{21}I_1 + h_{22}V_2$$

or

$$V_1 = h_i I_1 + h_r V_2$$
$$I_2 = h_f I_1 + h_o V_2$$

where i = input, r = reverse, f = forward and o = output.
Hence

$$h_{11} = h_i = V_1/I_1 \quad \text{with} \quad V_2 = 0$$
$$h_{12} = h_r = V_1/V_2 \quad \text{with} \quad I_1 = 0$$
$$h_{21} = h_f = I_2/I_1 \quad \text{with} \quad V_2 = 0$$
$$h_{22} = h_o = I_2/V_2 \quad \text{with} \quad I_1 = 0$$

Index